THE STORM OF THE CENTURY

THE
STORM
OF THE
CENTURY

TRAGEDY, HEROISM, SURVIVAL, AND THE EPIC TRUE
STORY OF AMERICA'S DEADLIEST NATURAL DISASTER:
THE GREAT GULF HURRICANE OF 1900

AL ROKER

wm
WILLIAM MORROW
An Imprint of HarperCollins*Publishers*

HarperCollins books may be purchased for educational, business, or sales promotional use. For information please e-mail the Special Mar-kets Department at SPsales@harpercollins.com.

A hardcover edition of this book was published in 2015 by William Morrow, an imprint of HarperCollins Publishers.

FIRST WILLIAM MORROW PAPERBACK EDITION PUBLISHED 2016.

Designed by Jamie Lynn Kerner

Library of Congress Cataloging-in-Publication Data has been applied for.

ISBN 978-0-06-236466-1

16 17 18 19 20   ov/RRD   10 9 8 7 6 5 4 3 2 1

CONTENTS

UNDERWATER

A MAN PINNED UNDER THE WATER STRUGGLES TO FREE HIM-self. Fifteen feet below the water's surface and the air he needs so badly, his thrashing body begins to weaken.

Big timbers, mercilessly heavy—only moments ago, they held up his house—are pressing him down. They hold him deep underwater. He can't move them. He begins to drown.

Knowing he has no chance, he stops struggling. The man tries to accept his fate.

Yet somehow when he comes to himself again, he's risen from the depths and broken the water's surface. Bobbing and kick-ing in a violently churning sea, he gasps for breath in the dark-ness, pelted by rain in a wind even louder than his gasps.

Two of the great house timbers that held him down have risen with him. They press him, one from each side. He clings to them.

He's alive. It's the night of September 8, 1900. And this is—or was—Galveston, Texas.

The man is Isaac Cline. Gripping the timbers that nearly

killed him, he surfs and circles randomly through a chaotic scene lit by flashes of lighting. There is no city. There is only this sea, its waves rising and falling as rooftops and treetops collapse into them. Objects of all kinds shoot from the swirling water and through the screaming wind: huge pieces of roof, doors, beds.

At every lightning flash, Cline scans the water. His pregnant wife and three daughters were with him in their second-story bedroom. He ducks projectiles in the air, dodges debris in the surf, tries with all his strength to stay above the rising water. Desperately seeking any sign of his beloved family, he fears the worst.

Isaac Cline knows exactly what's happening. He's a meteorologist—a weatherman—and one of the country's best. As he fights the wind, the water, and the flying debris, he understands that what's taking place tonight is something that he and other meteorologists of the United States Weather Bureau have long been certain could never occur here.

A hurricane, Cline has reassured the public, can't hurt Galveston, Texas.

Now Cline knows that this is a monstrous hurricane, and it has destroyed his city in only a few hours. He has been as wrong as it's possible to be. All of the systems, all of the miracles enabled by the modern science of meteorology, have failed.

There's no going back now. No way to correct the error. Galveston is gone.

What Isaac Cline can't know is that, more than a century later, this storm will remain not just the worst hurricane, but the worst natural disaster of any kind, ever to hit the United States:

10,000 or more lives lost in one night; higher winds and lower pressure than any previously recorded; damage estimated at nearly $20 million (more than $700 million in twenty-first-century money); a great city reduced overnight to miles of rubble. Adrift tonight on a surreal ocean of chaos where a city used to be, Cline can only go on scanning the dark maelstrom.

He can only pray for any glimpse of his wife and children. The best weatherman in America can only wish things had gone differently.

PART I

THEY ALL HAD PLANS

CHAPTER I

LOOKING FORWARD

A SCORCHING END TO A HOT SUMMER: THAT'S WHAT EVERY-body was saying. It was the first week of September in the grand turn-of-the-century year of 1900, and people in Galveston, Texas, were complaining about the heat.

That's one reason little Mary Louise Bristol, seven years old, was looking forward to the weekend. At her mother Cassie Bristol's boardinghouse, not far from the harbor on Galveston Bay, work went on despite the heat, just as it always did—and just as work went on everywhere else all the time in bustling, steaming Galveston. This island city in the Gulf of Mexico, lying two miles off the Texas mainland, had become the most important port in the state, the foremost cotton-trading center in the world.

Gangs of hustling longshoremen, black, white, and Mexican, loaded and unloaded ships on the wharves on the bay. Bankers and lawyers made deals in their offices above the broad sidewalks of Avenue B, known as the Strand. Cassie Bristol, a widowed mother of four, ran her boardinghouse near the harbor in hopes of feeding and bettering her family. Along with nearly 40,000 others, they all played their parts in a booming city's busy life.

Mary Louise Bristol, known as Louise, was Cassie's youngest. And Louise was the only member of her family who had any time for play. Sometimes she played alone. Sometimes she played with her friend Martha, who lived across the street.

But everybody else in the Bristol family was always busy with work. The girl knew her house wasn't only a home but also a living, and she knew her mother was a smart woman. Louise's father, a seaman, had died at sea when Louise was just a baby, and Cassie, left alone with four children, rolled up her sleeves, took out a mortgage on the house, added a second floor, and started renting rooms. Cassie had a goal: keep her children from falling into poverty and disgrace while teaching them the ways of a genteel life.

That ambition made Cassie's life a struggle. But it was a battle she was winning, thanks to constant work. Louise's brothers were old enough to have jobs and to bring home their pay. Her sister Lois, fifteen, helped their mother at the boardinghouse.

And the Bristol place was especially busy this week. School was back in session after the summer—Louise had just nervously started first grade on Monday—and every September, the house began filling to the brim with students of the University of Texas Medical School here in Galveston. Those young men were among Cassie's most reliable customers. Soon they'd be swarming over the place, their trunks hauled in by horse-drawn wagon and humped inside by porters, their rooms assigned, their questions answered, their beds made, their food served. Louise's mother was spending this week getting the place ready: cleaning and airing and washing and drying. Cooking and canning and stocking up on stores.

What the little girl most looked forward to—as the endless work went on at home, and the heat of early September remained so oppressive—was the weekend. Hoping to forget about school, hoping for a change in the weather, Louise was looking forward to Saturday, when she could play.

Saturday, of course, would come. When it did, on the eighth of September, Louise, her mother, and the rest of the Bristol family would find themselves fighting gigantic volumes of ocean and wind beyond anything they could ever have imagined. Saturday would plunge them all, with all of their fellow Galvestonians, into a dark world of sheer horror.

Arnold Wolfram too was wilting out the end of summer with his fellow citizens in that first week of September. Wolfram would have seemed a staid enough, ordinary enough Galvestonian—a member of the thriving German American community in that polyglot gulf town. At forty-three, Wolfram worked in a fruit and produce store, where he made sales and took inventory. He was married to the former Mary Schmidt. They had six children, with four still living at home.

There was nothing out of the ordinary about Arnold. And yet while many of Galveston's German Americans had come directly from Germany, Arnold hailed from another, even older German community: that of Philadelphia. Arnold and his brother Henry had grown up bilingual, attending Philadelphia's German-speaking schools. In young adulthood, Arnold had taken a factory job in his hometown, working for a hat company.

But Arnold and Henry found themselves drawn to the American West. Many young men born in the nineteenth century fantasized about seeking their fortunes there. Many found the idea impractical in the end or simply lacked the courage to make the move.

Not so the Wolfram brothers. In 1876, with the Civil War well over (they'd both been too young to serve), Arnold and Henry left Philadelphia and headed for the wide-open spaces of Texas.

The great cattle drives were peaking when the Wolfram boys arrived, and Texas ranches needed hands; the brothers found work quickly. Signing on at a ranch near Corpus Christi, Arnold Wolfram, the young German American romantic, was finally living the cowboy life. And it was at the fandangos— ranch parties held by cowboys and hands—that he met the famous Texas outlaws. They would drop in to drink and carouse. They were treated like ordinary partygoers. Arnold got to rub shoulders with them.

So now, almost twenty-five years later, this ordinary city groceryman Arnold Wolfram had something of a past. He'd once been a bit of a rambler, a rover, and a rider. It was said he'd even traveled with the fabled Texas Rangers. Brother Henry, more of a name-dropper than Arnold, claimed he'd met that most famous of Texas outlaws, John Wesley Hardin, that he'd been pals with celebrity Texas Rangers like Jack Helm and Sergeant Rudds.

Arnold himself remained more reticent about giving his former associates' names. But he did speak now and then about the outlaws, and about counting the Rangers as friends. And sometimes Arnold alluded mysteriously to a time when he had to ride bareback for the government on important official business.

Now a new century was beginning, and the cowboy days of cattle-boom Texas were fading into the past. Arnold Wolfram was a settled middle-aged man. He raised his big family, did a day's work for a day's pay, walked or trolleyed to work on Galveston's broad streets, and complained about the heat like everybody else.

Arnold Wolfram had good reason to believe that the wild times and big adventures were all behind him now. He was wrong.

Out on the east end of town, right down by the beach and the gulf waters, lived Annie McCullough. She was twenty-two, only recently married to Ed McCullough. Annie and Ed owned a little corner house with a sweet flower garden that featured Annie's prized rosebushes.

The McCulloughs' lot lay two short blocks from the gulf beach, almost perfectly level with the flat sand. Nothing blocked the view down to the water, nothing obscured the big gulf sky. The neighborhood was nice and, as a young wife, Annie McCullough was having a busy and happy time there. Ed was a hardworking and competent man, handling multiple jobs. Among them was making deliveries on his flat two-wheeled mule-drawn wagon, called a dray.

While she was not yet a mother, Annie nevertheless had a big family. She was a Smizer, one of the oldest African American families on the island, descended from the Galveston slaves who had been the first in all of Texas to receive news of their emancipation after the Union victory in the Civil War. The McCulloughs too had a history in Galveston's black community. Ed's relatives lived in nearby streets; a nephew of his lived with Ed and Annie. Annie's mother lived nearby too, as did many others in the Smizer family.

Annie's father, Fleming Smizer, however, worked for the federal government at the Custom House at Sabine Pass, an inlet controlling access to the city of Port Arthur, northeast of Galveston. So Mr. Smizer was often out of town, dependent on tugboats and other transport for periodically getting back to his family in Galveston. Ordinarily that wasn't a problem. He came home often.

The McCulloughs' east-end neighborhood on the beach was busy and thriving, but the lure of "beachfront property" had little meaning for Galvestonians in 1900, and the community on the gulf was largely working and middle class. The city's richest people lived some way off the beach. The really fancy homes were on Avenue J, also known as Broadway. Running parallel to the gulf and the bay, bisecting the city and itself divided by a nicely planted median strip, Broadway was Galveston's high ground—meaning it stood a few feet above sea level. The street was broad and elegant, with a trolley running down its median, and the palaces of the city's first families lined it on both sides. Those houses—rambling, with gables, porches, and porticos—combined the crazy excesses of ornate Gothic detail with a tropical mood left over from the island's French and Spanish days.

Where the newlywed McCulloughs lived, things were less fancy. This week, the main things on Annie McCullough's mind were her beloved roses, various family matters, and a new pair of shoes that didn't fit. She planned to have Ed return the shoes on Friday.

Newer to town than the McCullough and Smizer families were the Ketchums. Yet despite his standing as an outsider, Edwin N. Ketchum served as Galveston's chief of police.

In this first week of September 1900, Chief Ketchum was just getting back to town from a trip to Chicago, where he'd joined a reunion encampment of the Grand Army of the Republic—the organization for Union veterans of the Civil War. Ed Ketchum, top law-enforcement officer in Galveston, Texas, was a dyed-in-the-wool, unapologetic Yankee and former soldier of the Union army.

Galveston was full of proud former Confederate soldiers. Old and getting older, in 1900 they still held yearly civic celebrations to commemorate their role in the Civil War. They recalled the lost cause of Southern secession with greater sentiment as every passing year made that cause more remote.

So it was an indication of Galveston's unusually mixed culture that in 1900 the city accepted not merely as a citizen, but also as its most important policeman, a person who had come of age while waging war on that same Confederacy. Ed Ketchum's commitment to the Union cause was hardly subtle. He'd joined up as a drummer boy. By the war's end, he was a captain. He was proud of his service, and his northern reunion trips proved it.

But Ed Ketchum was not about to keep fighting the Civil War. That's something the city of Galveston, as a whole, didn't do either. Fifty-seven now, Ed was tall and skinny, a calm and genial man with a wife and eight children. The Ketchums lived in a solid, generously proportioned wood-frame house that had been built by one of Galveston's founders, Michel Menard, who had come from Montreal. Despite its recent connection to the Confederacy, by 1900 Galveston in particular, and the state of Texas as a whole, were products of recent immigration and settlement from far-flung places. Both the booming city and the booming state of which it was such an important part had been ethnically, politically, and racially mixed from day one.

And so Chief Ketchum, though a proud Yankee veteran,

was a popular man in town. On his grand lawn, he held annual picnics open to the public. The Ketchums were known to own one of the city's largest coffee urns, and when the Confederate reunion groups convened in Galveston, those aging Johnny Rebs borrowed the coffee urn from Ed Ketchum, the aging Billy Yank.

That week, returning from his Chicago jaunt, Ed confronted a pile of paperwork on his desk at City Hall—a grandly turreted stone castle, chateau-like, with peaked cupolas and a clock tower. Heat or no heat, there was nothing for Ed to do but get on it. Like everybody else's, Chief Ketchum's work went on that week.

And yet by Friday afternoon, when water from the Gulf of Mexico started running from the beach into town, and most of the citizenry of Galveston was preparing for nothing more than an exciting lark, Ed Ketchum would be among the first to start worrying.

It wasn't just the heat that first week of September. There was a stillness too.

Everybody could feel it, and Daisy Thorne, bicycling on the sidewalks of Galveston, was no exception. Daisy embodied the new ideal of modern American beauty—chastely appealing yet mildly athletic, forward looking yet demure. A schoolteacher of twenty-three, notably pretty, with luxuriant reddish hair, Daisy owned the first pneumatic-tired bicycle in town.

Cycling around town and along the beachfront, she wore a long skirt, a white shirtwaist, and a straw hat; a veil protected her face from the sun. To resist the sun still further, she bathed in the gulf waters only in the early evening, fully draped in the heavy bathing costume of her day.

Daisy was modest and genteel, yet she was also posing as a woodland nymph for an amateur painter named Mrs. McCauley. The young woman's physique and style—five-foot-four and 113 pounds, reddish hair pulled loosely back—offered just the kind of romanticism that painters and early photographers in 1900 loved to capture.

This adventure as a model verged on the risqué, and Daisy's mother had exacted a promise from the painter not to render her daughter's face in a recognizable way.

In real life, Daisy was a hard and focused worker. A graduate of the teaching program of the Sam Houston Normal Institute on the mainland in Huntsville—a three-year course that Daisy had completed in only one year—she now taught history, literature, and drawing to seventh graders at the Rosenberg Free School. She lived in Lucas Terrace, an upscale, fairly new apartment complex on Broadway's far eastern end—the same end of town as Annie McCullough, though slightly farther from the gulf beach. There Daisy shared a large two-story apartment with her widowed mother, her aunt, her sister, and her brother.

Like many genteel young women of modestly prosperous circumstances in 1900, Daisy spent her time at home in the pursuits of young womanhood considered appropriate in the day. She was an expert seamstress, sewing her own and others' clothes. She and her mother cooked the family's meals on the apartment's wood-burning stove. She liked to play the parlor piano.

But all that was about to end. Daisy had been engaged for three years, and at last she was to be married the next June. Daisy had met Mr. Joe Gilbert of Austin, Texas, at a sailing party on Galveston Bay when she was still studying in Huntsville. Joe had been a medical student then. Now he was a doctor.

At the party, Daisy had needed his steadying hand to board the rocking boat. Later she got ice cream on his pants, and from there the romance bloomed. It had grown fonder during the long engagement. Joe sent Daisy new novels and collections of poetry from medical school; by now, the books filled the apartment at Lucas Terrace.

And now that Joe Gilbert was established as a doctor, he and Daisy could marry. That meant Daisy would leave her mother's apartment. It also meant quitting work. Daisy was about to become a doctor's wife, and as the school year began, she was keenly aware that her professional life must end with this school year's conclusion. She savored the final days of summer, and of her single life. She knew everything was about to change. She had no idea how much.

Though a well-known citizen of Galveston, Boyer Gonzales wasn't home when the summer of 1900 came to an end. He was spending the summer and early fall in far-off New England. Boyer was fired by ambition to be a painter, and the northern seacoast offered qualities of light and color far different from what he'd found, and tried to bring to life in paint, amid the tropical scenes of his native Texas Gulf.

In fact, Boyer Gonzales was spending part of this summer with one of the greatest and most successful American painters of the day: Winslow Homer. The elder artist had taken up the younger one, and the two spent much happy time painting together at Homer's home in Prouts Neck, Maine.

If only Boyer could have spent all his time painting with Homer. But his artistic ambitions had never fit his parents' plans for him. They'd both recently died, and Boyer felt himself under even greater pressure to continue their legacy in Galveston. That pressure was starting to wear him down.

He was the fourth child and the second son of Thomas Gonzales and Edith Boyer, who lived in one of Galveston's more staggering mansions. Their importance was notable in part because the Gonzaleses were one of the few socially prominent Galveston families whose head was Mexican. Thomas, Boyer's father, had been born Tomás, in Tampico, Mexico, and had grown up partly in New Orleans in the home of a rich cotton broker, his elder brother-in-law. This rapidly assimilating youngster had gone to school not only in New Orleans but also up the Mississippi River, in Alton, Illinois. He had worked in the family cotton brokerage, then spent three years at school in Spain; at only fifteen, Thomas Gonzales had begun overseeing the family business office in Port Lavaca, Texas.

Thomas had started early, and he never stopped. Back in New Orleans, he soon married Edith Boyer of Philadelphia. Arriving in Galveston with his East Coast wife, he opened a wholesale grocery and cotton-factoring firm on the Strand. When he and Edith joined Trinity Episcopal Church, Thomas's assimilation was complete.

Thomas was conservative—a man of the receding nineteenth century. After serving with distinction in the Confederate army, he opened his own cotton firm. Contracts with textile factories in Europe made it one of the biggest shippers in Galveston's port.

He made a point of calling the firm Thomas Gonzales and Sons. The plan was for both Boyer and his brothers to join Thomas in the business. What other life could they possibly hope for?

Boyer did try. Raised amid immense privilege, the sensitive youth was sent to school at Williston Seminary in Massachusetts. His summers were often spent in cooling locales like Michigan and Maine. Back home, he became a young man about town, attending afternoon garden parties at the big homes and evening dances at the Garten Verein, a fancy, octagonal pavilion where the city's young elite held sway.

But at an early age, Boyer was also working hard in the offices of the family cotton brokerage. He was following the plan laid out for him. He was meant to reap the benefits of hard work and grand privilege.

So while Boyer Gonzales dreamed only of painting, in fact he'd become a family scion and a committed businessman who painted on the side. He attempted to mix the two. Hunting trips to the west end of Galveston Island—a sport of many upscale south Texans—had first inspired him to sketch and paint. The cotton brokerage had sent him to Seattle, San Francisco, and Boston, where he'd tried to capture on canvas the varying marine light of those far-flung places. Meanwhile, at his father's behest, he carried on the business.

But his heart was never in it. This conflict between his parents' plans and his own desires put Boyer under such stress that he developed a recurring respiratory problem. Seeking help, he started making frequent trips to John Harvey Kellogg's sanitarium in Battle Creek, Michigan. There he submitted himself to a stringent vegetarian diet, pioneered by the Seventh-day Adventist John Harvey Kellogg (it included the toasted corn flakes that would make Kellogg a household name). And yet, with each return to Galveston, and to the family business, Boyer's breathing problems returned as well.

Meanwhile, as Thomas Gonzales aged, that stern Mexican-American Texas war hero and Episcopalian pillar of Galveston society found himself dismayed by what he saw as a general downturn in values and morals among the youth. The onrush of the modern age did not suit him. The last thing Thomas could imagine was a Gonzales son taking up a loose bohemian profession like painting.

As Thomas grew older, he expected to give less day-to-day attention to the business. Boyer's elder brother, who actually enjoyed commerce, had died, and it was left to Boyer now, aided by the youngest brother, Alcie, to sustain both the brokerage and their father's commitment to the values and styles of late-Victorian Texas.

So Boyer's parents' deaths, coming close together, threw the artist into a kind of crisis. He inherited not only much responsibility but also much money, and he began using that money to fund painting trips. This summer of 1900, he'd been studying in Boston with Walter Lansil, the famous colorist, before joining Homer at Prouts Neck. These pleasures were guilty ones, stolen from his responsibilities.

And he'd been corresponding all summer from New England with Nell Hertford, another Galvestonian, who was at home. Nell and Boyer had known one another for years. He was her frequent escort at garden parties and dances, and friends automatically paired them, treating the couple as if they were nearly engaged.

Nell very much wished they were engaged. She was as cheerful and optimistic as Boyer was brooding and tortured. Boyer, however, remained remote, and Nell seemed to understand him. At thirty-seven, Boyer was depressed, emotionally paralyzed.

Nell contented herself this summer with writing him long, upbeat, newsy letters from Galveston. She hoped to keep up his spirits. Despite their long companionship, she still addressed him as "Mr. Gonzales."

Nell didn't know that, for Boyer, 1900 had already been a

year of change. He was trying to face up to his deepest ambitions as an artist. As another kind of trouble began brewing for his city, Boyer Gonzales's inner conflict was coming to a head.

Little Louise Bristol, looking forward to some time off from first grade . . . Annie McCullough, rose gardener, newly married to hardworking Ed . . . Arnold Wolfram, the grocer with a cowboy past . . . Daisy Thorne, schoolteacher, cycling enthusiast, soon to be married to Dr. Ed . . . Police Chief Ketchum, the genial former Yankee soldier . . . Boyer Gonzales, painting and brooding, far away in New England . . .

These and so many other citizens of Galveston, suffering end-of-summer humidity and stillness, could sense no impending calamity, of course. Their lives, and the lives of others in the horrific drama that began on Friday, September 7, were ordinary in that like all of us, they didn't see their lives and their families and friendships as ordinary but special. Life in Galveston that summer, like life everywhere, was both ordinary and extraordinary.

Out at St. Mary's Orphanage, run by the Sisters of Charity, for example—a big complex of stone buildings standing directly on the gulf beach just east of town—the lives of the residents might not have been considered ordinary by non-orphaned kids. But that place was the orphans' home. The ten nuns who took care of them served as both their family and their teachers. The Mother Superior, Camillus Tracy, was truly a kind of mother, both to the orphans and to the sisters she supervised.

So the orphans' lives went on that week, and the nuns' lives went on, as their lives always had. They looked forward.

Clarence Howth, a young lawyer with a wife, a new baby,

and a big, solid house only three blocks from the gulf, remained the supremely confident young man he'd always been. Not much bothered Clarence. With others of his type, he ate big lunches that week in the rowdy, cigar-smoke-filled environs of Ritter's Saloon on the Strand, exchanging the usual jibes and gossip. Like most of us, Clarence Howth didn't spend time imagining what it might be like to see his wife's lace curtains and wool rugs collapsing into an ocean that appeared almost out of nowhere. Like little Louise looking forward to Saturday and a day off from school, like Annie running her household and tending her roses, Clarence lived his life and pursued his plans.

Arnold Wolfram, long since pacified to a middle-class urban life, went to work at the grocery. Daisy planned her wedding to Dr. Joe, cycled, mused over the changes coming in her life. The police chief confronted the work that had piled up on his desk while he'd been away. Annie McCullough's father, Fleming Smizer, manning the customs post on the mainland, fully expected to return to Galveston at will, by ferry or tug. The Sisters of Charity taught and cared for the children as they listened to the sound, usually so mild, of steady gulf surf. They all had plans.

CHAPTER 2

THE STORM: AFRICA

Meanwhile something else was happening, so far away from the citizens of Galveston and their normal concerns that they couldn't possibly have imagined how thoroughly it would overturn their lives and plans. We know now that a hurricane arrived in Galveston at the end of the first week of September in 1900, and that its effects on Galveston were such that the hurricane continues to rank as the deadliest natural disaster in U.S. history. But in 1900, the origins of that hurricane would have been entirely invisible to the people it would hurt the most.

Even today, the exact origins of the Great Gulf Hurricane of 1900 remain a matter of conjecture. Most of the storms that the tropical regions of the globe are so good at producing never develop into hurricanes. Even those that do become hurricanes rarely inflict anything like the kind of damage that Galveston would experience that fall.

Nowadays, with modern forecasting technology—a fine-tuned integration of radar, satellite, globally networked communications, and digital imaging—each tropical storm can be watched before petering out, as most of them do. The storms that don't die out—the ones that grow, accelerate, and travel great distances—get tracked.

In 1900, however, tropical storms couldn't get close, coordinated scrutiny as they appeared, exploded, rained, thundered, traveled, and then, usually, died—or, unusually, became a hurricane. People in 1900 started watching storms only after they'd turned into something to watch. They could forecast: some weathermen got very good at knowing when a distant hurricane was coming. But being certain exactly where and when a particular hurricane had begun was more difficult.

By 1900, a deadly storm's likely origins could sometimes be determined in retrospect, thanks to observations made aboard ships near storms, or by those right in the middle of them. Those shipboard observations were recorded in logs and then communicated, when possible, to other ships by semaphore. In 1900, wireless ship-to-ship and ship-to-shore telegraphy was still experimental; weather reports made from sea could only be telegraphed on land, well after the observation itself had been made, and sometimes much later.

Meanwhile, the storm itself would be moving over great expanses of ocean. That meant that until a storm made landfall, it was moving faster than ships at sea could communicate with people on land.

Ship reports from the summer of 1900, studied today, suggest that the hurricane that made its most destructive landfall on the Texas barrier island early that September had its birth in a storm that at first looked nothing if not typical—and therefore, likely to be harmless. In late August, in the central Atlantic Ocean, a few hundred miles from the Cape Verde Islands, a series of events began to occur. Common events—ones that don't normally mean human disaster.

The Cape Verde Islands lie right off the big upper bulge of West Africa. If you draw a line on a map nearly straight westward from Cape Verde, over almost nothing but wide-open ocean, you arrive at Puerto Rico, gateway to the Caribbean Sea. Behind that island lie Cuba, the Gulf of Mexico, the Florida Straits, and the southern coastline of the United States.

Cape Verde is probably where the Great Gulf Hurricane of 1900 began. And that line on a map most likely represents its rough path. The Texas coastline is certainly where it went, and where it wreaked its worst havoc.

The area around the Cape Verde Islands serves as a kind of incubator for many of the hurricanes that do so much damage to the North American continent. Meteorologists have even enshrined the islands in a storm classification, the "Cape Verde Type Hurricane." That's because in summer there's an enormous volume of hot open water there, conducive to storm formation.

Yet because most such storms dissipate over that warm sea, the real question for researchers has long been not why some Cape Verde events become damaging storms, but why so many more don't. Every summer and fall—today, as in 1900—conditions there seem ripe for almost constant disaster.

What was happening in the late summer of 1900 off the shores of West Africa involved some of the most titanic forces on Earth. Citizens of Galveston, Texas, from Annie McCullough to Nell Herford to Arnold Wolfram, were destined to bear the brunt of big meteorological events they never could have imagined.

Around the continent's big, western bulge, mainland Africa gets most of its annual rainfall all at once: that's a classic rainy season. The region has what's called a savannah climate: rolling grasslands, with not enough annual rainfall to support big trees. Every morning, all summer long, as sunlight begins hitting the continent and moving across it, masses of hot, wet air start lifting from all that grass, from the soil, from the warm rivers, from the laboring people and the grazing animals.

The sheer size of this daily upward movement of water is astonishing. The air mass that results may be thousands of miles long. It can be many miles thick. And this happens every morning.

Every midday, this huge mass of wet heat has risen into a higher zone, where the air is cooler. The warm mass, less dense than the colder air up top, infiltrates the coolness and, for a while, it just keeps rising. But now as it rises, it changes: the surrounding coolness makes the hot air mass start condensing.

The wet air mass begins releasing a hot vapor. The vapor drives what is now a gigantic cloud of moisture still higher into the cooler air.

At last, the cloud has reached the highest point of cold air; it can rise no longer. The two air masses—hot and cold—are in conflict. As energy from the two conflicting air masses is released, thunder rolls.

The hot-air cloud starts to tower. It flattens out along the bottom. Meanwhile, water in the cloud begins to crystalize. Amazingly enough, given the heat below, it has soon frozen into icy droplets.

These billons of droplets start doing what they're named for. They drop. And as they drop, they melt, becoming rain. So all afternoon, every day, what was once that miles-thick mass of hot air falls back to the earth in the form of a hard rain, pounding the whole savannah. Under thick, dark, banging clouds, water deluges the grass and soil.

Near the ground, the temperature plummets. Rivers, swelling and flooding, pour silt onto the land. The silt promotes both crop and natural growth. The rainy season is punishing, but it is necessary to life.

By evening, most of the water collected from the ground in the morning has fallen as afternoon rain. Often the sky clears for a brilliant sunset.

The savannah rainy season can be dangerous, but because it's necessary, the people who live by seasonal flooding have found many ways of coping with it over thousands of years.

In the police station back in Galveston, Ed Ketchum would have complained genially of the summer heat, but he wouldn't have considered in any detail how other heat, in faraway places like western Africa, might interact with conditions in the Gulf of Mexico to bring disaster into his personal and his professional life. Weather science and weather mapping didn't bring information like that to Ed.

Sister Elizabeth Ryan and Mother Superior Camillus Tracy, caring for the orphans in the big building on Galveston's beach, had many things on their minds. The African savannah rainy season, and its possible effect on those orphans, wouldn't have been among them.

Even to people living on the savannah, pounded by flooding rains they equally feared and celebrated, the storms drenching their land that summer seemed to stay put. The rain came every day, just as it always had. Rain seemed to exert its fierce effort solely on the ground it rose from and was busy enriching.

But to people in ships at sea, far from the original sites of the African rains, the savannah's rainy season had other visible effects. Ships' captains were among the few people on Earth capable of observing directly how those storms moved, and the violence that their path, very occasionally, could cause. Late in August of 1900, one such movement began to occur.

And here another titanic planetary force enters the hurricane equation: wind.

Like the rains, many winds are seasonal, and people have been giving those winds names since before anyone can remember. *Meltemi,* blowing north from Africa across Greece in late summer. *Abrolhos,* the spring and summer squalls off eastern Brazil. *Chinook,* the hot, dry hard wind that can blow in winter between the North American plains and the Rocky Mountains, melting snow. There are dozens of others.

Scientists too have names for seasonal wind patterns. Every January, for example, a wind pattern known as "the African easterly jet" blows, as its name suggests, from east to west across the African continent.

Effects of the African easterly jet are felt by people on the ground, of course. It's windy down there. But the real action isn't on the ground. This wind blows nearly 10,000 feet above sea level.

Early in the year, the jet is blowing over the fifth parallel— that's pretty far south, just above the equator. At that time of year, the jet's speed can get up to 25 miles per hour. Brisk, but nothing to write home about.

By spring, however, that same wind has moved northward and increased its speed. In summer, it's blowing stiffly straight across the mid-point of that high, bulging part of West Africa. And because it's summer, the African easterly jet starts interacting with the rainy season. Meanwhile, it reaches its highest speed: about 30 miles per hour.

That's nothing like gale force. Still, thanks to other conditions, it's enough to start a chain of events that, sooner or later, and very far to the west of where the jet is blowing, can produce winds of over 100 miles per hour, the kind of gusts that would help knock down Galveston, Texas.

What happens is this. The African easterly jet can push a batch of those daily, rainy-season thunderclouds northwestward, out over the Atlantic Ocean. There, the thunderclouds encounter another ocean wind, blowing from a different direction. The African easterly jet, curving northwestward, meets the northerly Atlantic jet, curving southeastward.

The two winds don't crash head-on. They're moving on curved paths, not racing straight at one another. They spin past each other, brushing hard as they try to pass in a narrow corridor.

The air in this spot—it's known today as the Intertropical Convergence Zone—thus develops into a kind of atmospheric fold. One effect of the fold is that the African jet, interrupted in its northwest trip, starts to bump and ripple—really, to wave. Its westward flow takes on a regular waving pattern.

That's one reason we call this whole oddball system—a gigantic African rainy-season storm cloud, swept out to sea and caught in a pocket of weirdly opposing winds—a tropical wave.

On Monday, August 27, 1900, a tropical wave appeared over open waters about midway between Cape Verde and the Antilles. We know that because a captain directing his ship through those waters made a log note on it. Unsettled weather, the captain recorded. Force Four winds, out of the east—only 13 to 18 miles per hour.

Rain and wind in a summer sea: nothing for an experienced captain to worry about. Nothing to semaphore, no need to raise storm-warning flags for other ships passing that way.

The captain saw conditions like this all the time.

And yet that one captain's note represents the first recorded sighting of the system that had begun as a tropical wave off Africa to the east and had started spinning. This was the system that, out of so many other storms, would blossom into the worst natural disaster the United States has ever experienced.

The next day, another ship also made a note: stormy weather, winds from the south-southwest, at Force Six, maybe 30 miles per hour. Nothing like a hurricane.

No further sightings of that storm were recorded until it had traveled thousands of miles and arrived in the Caribbean Sea. Then, as the storm approached Cuba, thousands of miles from where it began, Cuban meteorologists, among the best in the world at forecasting, began to track it. And they began to worry.

That was on Monday, September 1. Unobserved, the system had been moving fast across the ocean. And it had already reached monstrous proportions.

CHAPTER 3

A REASONABLE ARGUMENT

In Galveston, all this talk of drenching rains and wildly conflicting winds, occurring over West Africa and then flying out over the sea, would have seemed remote at best. Galveston began that week with no wind. The sky was clear, the heat wave was holding on, and Africa was far away.

But even if the storm flags had been flying that week, many Galvestonians would have shown little concern. There had been storms here before—many of them, even big ones. They were nothing to worry about.

You weathered a storm, you picked up the mess, you moved on. The way Galvestonians looked it, this wasn't some primitive little village, like those of old Europe, built on earthquakes and cowering in fear of nature. This was 1900—the dawn of a new century. This was the United States of America.

This was Texas, the biggest state in a booming nation.

And this was not only Texas's greatest metropolis but also

one of the world's greatest ports. This was the rich and beautiful island city of Galveston, Texas, U.S.A.

At first, the place must have looked unlivable. A flat, sandy, narrow plain two miles off shore. Humid. Baking, mosquito-infested, in the gulf sun and salt. And no fresh water. To European eyes, there was nothing there.

Yet French and Spanish sails began appearing on the gulf horizon, and casing out the island, as early as the late 1500s. It was a key factor in the later development of Galveston—and in the evolution of its unique culture—that those ships didn't come mainly to establish civilian settlements on what would become the vastness of mainland Texas. That land was arid and forbidding. Along the coast, Karankawa, Caddo, and other indigenous people made European settlement on the mainland challenging; far inland, on great plains, the Comanche empire ruled and defended complex trading networks. At first, Europeans showed little interest in venturing into Texas.

Instead, the imperial ships came to the gulf mainly to establish, by their military presence, competing monarchs' claims on the New World. In hopes of enforcing supposed borders between European territories on the Gulf Coast—borders that existed more clearly on maps in far-off Versailles and Madrid than they ever did here on the ground—these ships brought more soldiers, adventurers, and priests than settlers.

And it would have seemed especially crazy to settle, of all places, on Snake Island, as the Karankawa people who lived and hunted there—and whose reputation for cannibalism terrified and fascinated the Europeans—called their long, narrow barrier. There were, in fact, snakes on that island. There wasn't much else. Really a glorified sandbar, the island ran exactly parallel to the slanting coastline, northeast to southwest. On its northeastern end, it looked across a small inlet toward the southwestern end of a long peninsula, one day to be named Bolivar, which came from farther northeast on the mainland. Together, those tips of land—the island and the peninsula—formed a gateway into a massive, sheltered bay, to be named Trinity, which opened deeply into the Texas mainland.

At its southwestern end, the narrow barrier island nearly reconnected with the mainland. The water bounded there by the island and the mainland gave the island its own smaller bay, a calm, hot place very near the open gulf. In the diplomatic and military struggle over the New World, that bay came to form one of the strategic harbors in what Spain considered one of its most important American possessions: Mexico.

The coastline of that great Spanish province Mexico soon stretched—according to Spain—from somewhere near today's Bolivar Peninsula to Corpus Christi, then to Matamoras, and then all the way down to the Yucatan. The northern interior of that territory became known as Texas. Though barely settled by the Spanish, Texas was nevertheless overseen and administered from the south by the Spanish colonial government in Mexico.

The status of Snake Island, however, long remained in doubt. The broad, tranquil bay between the barrier island and the mainland wasn't far from the area that France considered its own Gulf Coast, which lay to the east. The French Gulf was centered on New Orleans and Biloxi, two trading towns near the mouth of that great interior river the French controlled: the Mississippi. Snake Island thus served as a gateway of great importance between competing sides in an imperial contest. By the late 1600s, the Spanish were calling the island Isla Blanca. The French called it Saint Louis.

But which European power was actually in charge of this dry, barren strip of offshore land? For years, that question couldn't be answered. None of them were in charge, really.

And so, through decades of wild colonial turmoil, this long, thin island, commanding a big and beautiful bay, became the scene of a revolving series of inconclusive tropical adventures. They would mark its people and attitudes for good.

The island got its modern name in 1785, when a Spanish explorer called the bay, strangely enough, the Bay of Galveztowm—"towm" with an "m." That word mingles the name of the Spanish viceroy in Mexico, Bernardo de Gálvez y Madrid, with an oddly English-sounding word. It thus foreshadowed some dramatic changes, for both Galveston Island and the whole Gulf Coast.

For as the eighteenth century ended, a new entity—not European, not a colony, but both North American and independent—came into existence. It called itself the United States of America.

This ambitious new nation was looking far beyond the old borders that had restrained it when, as a collection of English colonies, it had stayed pinned to the Atlantic seaboard. The Americans were gazing eagerly westward, of course. But they also looked southwestward, toward the Gulf of Mexico. They even eyeballed the big interior regions, little known to them, that the gulf led to.

On the Gulf Coast itself, meanwhile, the political situation had grown even stranger. European sovereignty abruptly shifted, then abruptly shifted back. Those seismic movements added to the confusion. And they encouraged American interest.

In 1762, for example, France ceded to Spain, wholesale, the entire area it called Louisiana. That didn't mean the region later known as the state of Louisiana: it meant pretty much everything from the Mississippi to the Rockies. In the gulf, France handed over to Spain all of what had been French Biloxi, New

Orleans, and Baton Rouge. Shortly the French were gone, never to return.

But then, in 1800, Spain ceded all of that land back to France, and France, with no desire to use the property, flipped it, almost overnight. Napoleon Bonaparte marked up the American West, which he'd just received back from Spain, and sold it lock, stock, and barrel, all 828,000 square miles, to the United States in 1803.

In the United States that transaction became known as President Thomas Jefferson's Louisiana Purchase. Suddenly, along with much of what would become the great American West, a key city on the Gulf Coast came under U.S. control for the first time. New Orleans, with all its channels, bays, and bayous, gateway from the Mississippi River to the Gulf of Mexico, was now an American city and a U.S. port. For the restless, entrepreneurial American citizens who craved land, profit, and adventure, New Orleans couldn't have looked more exciting.

There was one problem, however. The vast region known as Texas had barely ever been settled by white people. It was therefore highly inviting to those same restless Americans. And now Texas lay conveniently near some new territory gained by the United States in the Louisiana Purchase.

Yet Texas remained part of Spanish Mexico. Never ceded back to France by Spain, it hadn't been transferred by France to the United States. A new border, not fully agreed upon, therefore came into being. The border separated Spanish Texas from the new territories of the United States, bounded to the south by the Gulf Coast. And the border made everybody uneasy.

In 1817, seven ships sailed into Galveston Bay. The captain of the fleet was the famous French pirate Jean Lafitte. He was looking for a new headquarters.

The island at the head of the bay seemed the ideal place, and Lafitte already knew it well. By now, he was feeling old. Though renowned as a pirate, Jean Lafitte was hardly the head of some ragtag band. A powerful military strategist, a connoisseur of fine food and wine, he'd trained under the great conqueror Napoleon Bonaparte. Then, as chief administrator of a complex smuggling network, based at first in New Orleans, Lafitte commanded an expert naval artillery battalion, at once feared and envied by all the competing colonial powers in the gulf.

Lafitte's operation ran illicit goods from the Caribbean colonies into New Orleans. The pirate violated the U.S. Embargo Act of 1807 to provide elegant residents of that city with the luxuries they loved. Using islands off New Orleans, Lafitte managed illegal, well-defended warehouses and wharves of his own, where he employed hundreds of men.

He faked ships' manifests and eluded customs. His fleets of privateers took gold from nearly every nation's ships. They took the ships too, and added them to their fleet. Lafitte also traded in enslaved Africans.

The captain wore black and was the author of his own myth: aristocratic parents guillotined in the French Revolution, a series of duels of honor, multiple common-law wives at once, a raconteur, a brooding and violent antihero. For a time he charmed New Orleans society even as he lived outside the law.

Underneath all that charm and style lived a shrewd tactician. By the time he took up residence on Galveston Island, Lafitte had been playing every side in the political turbulence on the Gulf Coast against every other side. Things here had now gone truly wild. Mexicans began rebelling against Spain in 1810. That brought freelance armies of Americans flooding into the area to fight on behalf of the Mexican revolutionaries. These improvising Americans, known as "filibusters" and often operating outside the sanction of U.S. law, hoped to gain land from an independent Mexico. Pirate organizations also took part in that revolution, mostly on behalf of the Mexicans.

Then the British showed up. In 1812, during the war named for that year, they established a base at Pensacola and began trying to wrest New Orleans from the United States.

All of that booming cannon fire and maneuvering under sail among a multitude of mutual enemies served Lafitte well. Wanted in the American territory of Louisiana for breaking U.S. laws, Lafitte nevertheless turned down a British offer of alliance in the War of 1812. He offered help to the United States.

General Andrew Jackson was deeply skeptical of Lafitte's trustworthiness. Yet as the commander of American forces against the British, he made a deal with the pirate. Jackson would pardon Lafitte's men and release Lafitte's ships captured by the United States in a recent raid; the gentleman pirate would, in turn, deploy his mighty naval organization against the British on behalf of the United States.

The famous Battle of New Orleans, the United States sent the British packing for the last time. Jean Lafitte's crack artillery, and his shrewd advice to Jackson on naval strategy, played crucial roles in the victory.

The dashing pirate did not rest. Lafitte next offered his services to Spain. He agreed to spy on the Mexicans who were rebelling against the empire. It was this scheme that first brought Lafitte to Galveston Island.

Another French pirate, Louis-Michel d'Aury, had made Galveston a base of operations, both for seizing ships and for supporting the Mexican revolutionaries. Lafitte went to the island at Spain's behest to gather intelligence on d'Aury's operation. But he really spent his time there sussing out the island as a future headquarters for his own smuggling business.

Soon Lafitte and his men were alone on Galveston. D'Aury had taken his men to fight for Mexico; when he returned to the island, he found Lafitte in command. Respecting his colleague's ruthlessness, d'Aury turned his fleet around and left for good. Lafitte ruled Galveston.

The island was now, according to Lafitte, an independent kingdom. He called it "Campeche," and the pirate himself was the head of its government. He established his own court of admiralty and began running the place as a benevolent dictatorship, devoted to smuggling.

He built himself a big, solid, red two-story house, La Maison Rouge, complete with a moat; it faced the harbor. But he spent most of his time on the *Pride,* his flagship anchored in the bay.

Lafitte's people built homes on the island. Over time they created a small town. The pirate king accepted newcomers on the condition that they swear loyalty to him. Soon he had built a community of 200 men and a number of women with allegiance only to Campeche, which meant to Lafitte himself.

Lording it over the Karankawa, who had used the island for generations, the citizens of this shadowy, self-declared outlaw state raided ships of the United States and every other nation in the Gulf of Mexico. These were the first full-time European residents of Galveston Island.

It's almost impossible to believe that fewer than eighty years passed between 1821, when Jean Lafitte—finally kicked out by the U.S. Navy—sailed away from Galveston and disappeared forever, and 1900, when Daisy Thorne, the cycling schoolteacher, Ed Ketchum, the Yankee police chief, the Bristol family, and so many others were making their plans and living their lives in a booming modern city.

Those eighty years could not have been more eventful, for Galveston, for Texas, and for the United States. The pirate king, somehow brought back from the dead, wouldn't have believed his eyes.

The startling changes that Galveston underwent in that brief span of time would continue to mark its people's moods and attitudes. For it's putting it mildly to say that Galvestonians became accustomed to coping with dramatic events. By 1900, a fort guarding Galveston harbor could boast of having been under the command of Spanish Texas, Mexican Texas, the independent Republic of Texas, the U.S. state of Texas, the Confederate state of Texas, and then the United States of America again. Galvestonians took it all in stride.

Things changed fast. First, the Mexican revolutionaries did, in the end, kick out the Spanish. One of the most important effects of that change was that American settlers started putting down deep roots in Texas that would prove fateful to the region's future and to the development of Galveston. The first deals for settling Americans in Texas, made by Moses Austin, a Missouri Territory mining entrepreneur, were originally struck with Spain, but when Mexico became independent, the new nation chose to honor them. In 1823, Mexico certified the American settlement, now led by Moses Austin's twenty-four-year-old son, Stephen.

Mexico quickly regretted it. These Americans were heedless of authority. They kept expanding to new territories, breaking the rules of settlement. They soon outnumbered Mexicans in Texas and seemed to be colonizing the whole place.

Some of the American settlers, harried by the Comanche, seemed bent on Indian eradication. Rekindling the old dreams of the filibusters, they imagined taking over Texas and extending their holdings all the way to the Pacific Ocean.

Settlers like Sam Houston harbored a different hope. Sam Houston wanted the Texan settlements annexed by the United States. Texas might then become an exciting new U.S. territory—even a new state.

Especially problematic for Mexico: these Americans came with enslaved Africans. The settlers saw Texas as an ideal place to continue and develop the South's "peculiar institution," just as its persistence in the South was beginning to roil the whole United States. The Texans' plans for big bonanzas depended on slave labor.

But Mexico objected to slaveholding. In 1823, the nation had banned slavery in its territories. Americans failed to comply, and the Mexican government soon decided it didn't want any more of these immigrants pouring across its borders and flooding its countryside with objectionable practices. In 1830, Mexico banned all Americans from entering Texas.

So it was that in 1835 war broke out between the Texan settlers under Stephen Austin and the Mexican colonial government under President General Antonio Lopez de Santa Anna. Famous scenes ensued. The best known is probably the Battle of the Alamo, in which all American fighters died. Among them were William Travis, James Bowie, and Davy Crockett.

The most consequential conflict in that war took place along the marshy western banks of Galveston Bay: the Battle of San Jacinto, where, in April of 1836, Texans serving under Sam Houston routed the Mexican army in less than twenty minutes. The independent Republic of Texas, founded by settlers from the United States, came into existence.

Thus the sovereignty of Texas—and of Galveston Island—switched yet again. Galveston was now subject to a freewheeling, self-created government like Jean Lafitte's. But unlike Lafitte's, this one was formed by ambitious Americans.

In April 1861, one week after federal ships fired on Fort Sumter in South Carolina, Sam Houston crossed Galveston Bay to the island, disembarked at the wharves, and angrily made his way on foot to the Tremont Hotel, one of Galveston's finer establishments. Houston's visit caused much public excitement. He wasn't merely the hero of the Battle of San Jacinto a quarter-century earlier. He'd been the president of the Republic of Texas, and then a governor of the U.S. state of Texas. And he'd just been shoved out of office for refusing to swear loyalty to the Confederacy.

The old man had come to Galveston to give a speech. He wanted to warn the people of Texas against seceding from the Union.

By now, Galveston was a busy port with a working customs house. A consortium of settlers and speculators, led by the French-Canadian Michel Menard, had begun building homes and businesses under the independent Texas republic, and the city of Galveston had been for a time the capital of that republic.

Still, as late as the 1840s, a visitor had described the city this way: "The appearance of Galveston from the harbor is singularly dreary . . . a piece of prairie that had quarreled with the mainland and dissolved partnership." The island had seemed nothing but a grim Texas frontier town located, strangely enough, in warm tropical waters.

But throughout the 1840s and 1850s, the city began booming. Immigrants seeking opportunity in the new republic came not only from the United States but also, crucially for Galveston, from abroad: Germans, Greeks, and Italians immigrated to Texas; Jews of all classes came from eastern Europe; and the city's topographical position in the gulf made it the region's major port. Galveston soon mingled Mexicans, Europeans, and American families, as well as African Americans both free and enslaved.

Catholic churches and convents were built. The formation of a multicultural port city in the Texas Gulf was well under way by the 1860s.

By the time Sam Houston went to the Tremont to give his speech, he knew had no chance of persuading the people of Galveston, much less all of Texas, to change their political course. But he was explosively irritated. From the earliest days of Texas immigration, Houston had favored not independent expansion into California but annexation by the United States. There, he knew, lay prosperity for Texas.

And his dream had been realized—if all too briefly. In December 1845, Texas was admitted into the Union. The former independent republic became the twenty-eighth American state.

But now the slavery issue had boiled over in the United States. And while the mixed nature of Galveston's population did make the city more progressive than the rest of Texas—Germans, Jews, and other European immigrants generally opposed slavery—and while slave labor wasn't playing an important role in the city's economy anyway, after the shelling of Fort Sumter, anti-Union hysteria ran amok on the island.

A mob ransacked offices of a Galveston anti-slavery German newspaper. Federal supplies were now in the hands of amped-up local militias. And as Sam Houston, a founder of Texas, walked the city's broad and busy streets, the people were actually shouting at him, catcalling and insulting him.

The old man needed the right place from which to make his speech. The city had developed markedly from Menard's first days: there was a railroad bridge to the mainland now; fine homes and public buildings represented the beginnings of the city's characteristic Victorian-tropical architecture. And, as in New Orleans, galleries covered many sidewalks to shade the passersby and provide upper balconies.

Finding himself locked out of the courthouse, Houston chose the Tremont's gallery for his speech. He stepped out onto that balcony. He glowered down at the angry crowd.

What Sam Houston told Galvestonians that day—"Will you now," he thundered, " . . . squander your political patrimony in riotous adventure, which I now tell you, and with something of a prophetic ken, will land you in fire and rivers of blood?"— turned out to be somewhat exaggerated for those who remained on the island itself. It is true, however, that Galveston didn't exactly benefit from its brief period as a Confederate city.

Confederate strategy largely ignored the island. Many citizens fled to the state's capital at Houston, on the mainland. When the local Confederate command, too, finally evacuated the island, Galveston was left defenseless. Union troops sailed into the harbor and were greeted warmly by Galveston's mayor. Most of the remaining citizens' hearts really weren't in resisting.

But Union forces, occupying Galveston during much of the war, also lacked support from their own high command. Local militias poked about the island at night, skirmishing with the occupiers.

Still, things on the island stayed more or less calm during the occupation. So peaceful, in fact, that the Union general didn't even bother destroying the railroad bridge to the Confederate-held mainland.

That was a mistake. In 1862, Major General John B. Magruder, newly appointed Confederate commandant of Texas, took Union occupation of Galveston as a personal affront and decided to liberate the city. On New Year's Day 1863, Magruder attacked the island by both sea and bridge. During the Battle of Galveston, he retook the city for the Confederacy.

Galvestonians might have treated this change in authority as yet another in the long series of upheavals that had made them who they were. But in taking the city, Magruder shattered the relative calm that had prevailed during the occupation. In the Battle of Galveston, hundreds of men on both sides were killed and maimed. The wounded filled the Ursuline Convent.

For the first time, Sam Houston's prophetic words seemed to be coming true. Most Galvestonians found all the violence and killing pointless and disgusting. They developed a "pox on both your houses" attitude regarding the opposing armies— and regarding the Civil War itself. That attitude would become an asset to Galveston's future growth.

Galveston got over the Civil War more quickly than did many other cities and regions of the former Confederacy. As the war wound down, Confederate troops mutinied against Magruder; at the bitter end, troops pillaged the quartermaster's store. For a time, anarchy threatened to prevail on the island.

But it was in the city of Galveston, on June 19, 1865, that Union authority first arrived in Texas, raised the Stars and Stripes, and took command of the whole state. On that day, and in that city—forever celebrated as "Juneteenth"—the abolition of slavery in Texas was first announced.

And the city on the narrow gulf island now began a rise to dominance, in Texas and throughout the world. That ascent may be unique in its speed and intensity.

In 1900, thirty-five years after the Stars and Stripes first flew again in Galveston, about 37,000 people lived in the city. That was fewer than lived in Houston, but size had never been Galveston's stock in trade. This was an island. There was no way to grow beyond certain obvious physical limits.

And yet by the last quarter of the nineteenth century, Galveston had electric lights on its broad and graceful streets—the first electric lights in Texas. Gaslight had shed a discreet, cozy glow; these new carbon-arc street lamps, outputting a blinding 4,000 candlepower, lacking the stink and soot of gas, turned the night into a white-out version of daytime.

And because in the 1880s Thomas Edison had suspended carbon filaments in vacuum-sealed bulbs, incandescence had tamed electric light—both its glare and its cost—for smaller spaces like warehouses, banks, and side streets.

Electricity was also powering another set of strung cables: the telegraph. Developed by Samuel Morse, the system had linked East and West on the eve of the Civil War, but by 1900 cable-laying ships plowed the seas as well. Ten days had once been the quickest time for a ship to bring a message from the United States to Europe. That gap had been shortened to seconds. It was rumored by 1900 that the telegraph would be going wireless.

Meanwhile a new device fostered yet another communication miracle. Voice-to-voice talk via the telephone was changing everything yet again. By 1900, switchboard exchanges could patch multiple calls quickly through a single system.

At first such systems were operated by boys. But it turned out that the anonymity involved in telephony tempted the boys to mouth off at callers. Soon grown women—those with an especially mature and smoothly gracious manner—replaced the kids as operators.

So Galveston had telephones and telegraph. The city's electricity, distributed by a turbine-driven central power system, flowed to public buildings and powered a reliable streetcar system. There were flush toilets. On Broadway the huge Gilded Age mansions of the city's industrialists and traders and financiers vied with one another for ostentation.

As a center of international trade, Galveston now epitomized the heady pace of growth and development that marked the entire country. This was the age of the robber barons: the great financiers, the railroad entrepreneurs, the steel magnates. This was the era of "the millionaire"—and Galveston, Texas, boasted more of them per capita than any other city in the United States.

There was conflict in America too. In the summer and fall of 1900, the incumbent president William McKinley and the Democratic populist William Jennings Bryan were restaging their epic political contest of 1896. Bryan came to Galveston that summer to give one of his powerfully emotional speeches.

The debates of the campaign season included such things as the goals of America's new ambitions abroad: President McKinley had sent U.S. troops to China, Cuba, and the Philippines. People were questioning the immense power of the high-finance elites. Some were championing the rights of labor.

Those themes would resonate, of course, in American political campaigns for decades to come. Amid all the frenetic activity at home and abroad, the nation was leading the world into new kinds of success, and new kinds of turmoil, which would come to mark the modern age. Galveston, Texas, had become one of the main centers of that turmoil, and that success.

So by 1900, some had begun to call Galveston "the New York of the Southwest." At the very least, this was another New Orleans, another San Francisco. The city was laid out on a stately rational grid. Numbered streets, leading from the busy port on the bay down to the beautiful beach on the gulf, and lettered avenues, running across town, were jammed with horse-drawn carts hauling supplies, with elegant buggies driven by coachmen, and with electric streetcars.

Visitors crossing the bay by steam or sail or on a railroad causeway were struck at first by the deep-water wharves, bustling with the steady loading of bale after bale of cotton—along with bushels of corn, barrels of flour, and bundles of sawn lumber—and the equally steady unloading, warehousing, and trucking away of everything from beet sugar to cement to coffee.

Two new forts stood guard over the strategic island. They were U.S. forts, of course—but their names recalled the fight against Mexico for Texas independence and the establishment of the Texas Republic. Fort San Jacinto, named for the victorious nearby battle that wrested Texas from Mexico and established in 1898, stood at the east end of the island, commanding the bay. Fort Crockett, named for a hero of the loss at the Alamo, was even newer—just starting construction, in fact, in the summer of 1900. There, just west of the city, the U.S. Army Coastal Artillery was creating a modern facility on the gulf that could command the entire area with the biggest modern guns.

Beachgoing visitors, traveling southward across town on a special mule-drawn trolley, would arrive at the gaudy beach bathhouses on the gulf. For by now, Galveston's beautiful gulf shore was attracting tourists. The Pavilion, built as a beach palace, and illuminated by electricity, had rivaled Brooklyn's fabled Luna Park on Coney Island; it boasted a huge dance floor. When it burned to the ground in 1882, it was replaced by a grand beach hotel with a high dome, breezy porches, and a freshwater fountain gracing its driveway.

The city was one of the fanciest and most elegant tourist destinations in the country. Bathers especially loved Galveston's tranquil, warm gulf waters.

Trade, of course, was the basis of this remarkable rise to national importance. Galveston's harbor, where a pirate flagship had stood at anchor less than eighty years before, now shipped more than two million bales of cotton annually, worth more than $85 million. The Galveston Cotton Mill and the city's Texas Star Flour Mill ran at full tilt.

A bell opened and closed business on the vast trading floor of the Cotton Exchange near the wharves. There, below high chandeliers dangling from a spectacularly soaring ceiling, the marble walls echoed with prices and bids for bales and futures shouted by bearded men in coats and derbies and top hats.

Galveston also boasted dockside warehouses full of goods from all over the world. Those goods moved to Houston, Dallas, the South, the West, and directly to New York by steam, by sailing ships, square-rigged to fly a multitude of sails, and by rail. In a city this physically intimate, the new opera house, built on a grand scale in 1894, stood not far from the bordellos: the city's rugged frontier origins survived its rise to elegance and power.

Galveston may have been the New York of the Southwest. But it was also the first city of a state that fully captured the new American spirit of 1900. With the restoration of U.S. government after the Civil War, Texas was enjoying a huge boom as a national leader in cattle ranching and lumber production. Combining disdain for Northeastern-style niceties with eagerness for a Wild West form of cosmopolitanism, Texas style matched the feverish development, big money, and gigantic ambition that had gripped the nation as a whole. By 1900, Texas was leading the nation in cotton production. The oil boom to come was presaged in 1898, when the Corsicana oil field was opened by a former Standard Oil executive. The first modern oil refinery west of the Mississippi, Corsicana augured a new age for both Texas and the United States.

So in Galveston, the state's greatest port, there were 500 saloons. And because the city's trade was international, so were its people. Multiculturalism had only intensified since the Civil War: blacks, white Protestants, Jews, Germans, Irish Catholics, and Latinos mingled with relative ease. By 1900, Galveston was the favorite destination for European immigrants entering the western United States. Relative racial and ethnic tolerance were characteristic not only of Galveston but also of post–Civil War Texas as a whole. By 1883, the Negro Longshoreman's Union, along with a black screwman's union, were admitted to the Galveston Trades Assembly.

Yet the nature of those unions shows how thoroughly racial segregation prevailed in key areas of Galveston life in 1900. Black and white people lived on some of the same streets but rarely socialized as equals. Schools were segregated: Galveston's Central High School was the first high school in Texas for black students. Yet early on, the city's school board determined that teachers and principals in those schools should be hired without regard for race. Many of the city's African Americans found avenues to middle-class life by serving as teachers and administrators in the "colored" schools. The quality of the education they provided was widely considered to be high (some thought higher than that provided white students). It contributed to the rise of the black community in Galveston.

Still, segregation would hamper that community's success for more than sixty years. Black citizens were expected to ride in the backs of trolleys. Theaters had separate "colored" sections. Swimming—or "bathing," as Daisy Thorne would have called it—was not racially integrated: a small section of beach was designated for Negro citizens.

And while the civic leadership and upper-crust society, based so entirely on success in trade, was unusually well mixed—Jews and Mexicans were included with the Anglo-Protestant elite—African Americans were not among them.

In 1900, those injustices, underlying Galveston's evident racial calm, would be exposed by catastrophe. The calm would be broken by a storm.

All this dazzling success had come about thanks to the amazing capacity of Galveston's people, and of its leadership, to solve problems. By the fall of 1900, as Galveston looked at its own situation, all of the biggest challenges seemed to have been met.

These were challenges posed by nature. In the 1880s, the city's business leaders had begun grappling with what had seemed Galveston's biggest challenge: the harbor, too shallow to accommodate the sheer size of the new steamships that were now hauling great volumes of freight about the country's harbors and across the seas. To rank as a first-class harbor, ships drawing twenty-six feet of water at low tide had to be able to navigate there. In 1875, a sandbar at the entrance to Galveston's harbor allowed only eight feet of clearance.

Worse: New Orleans was beginning to build jetties that might give it first-class-port levels of clearance. Houston had started dredging a channel that would tempt shippers to avoid Galveston altogether. In the 1800s, it looked as if nature—in the form of the tidal sand at the entrance to Galveston Bay—might stifle the island city's rise.

By 1900, however, that problem had been solved. Solving it took drastic action, intense political commitment, and ambitious engineering. At the Cotton Exchange, the city's leading businessmen took on nature. These were heads of the great families of Galveston, owners of astonishingly luxurious mansions on and around Broadway: the Sealys, the Moodys, the Blums, the Kempners, and others.

They formed the Deep Water Committee (DWC). Selling bonds, coaxing federal money from Congress, and even personally lobbying President Cleveland, the big men of Galveston brought in one of the great military engineers of the age, Colonel Henry Martyn Robert, of the elite Army Corps of Engineers. In the Civil War, Robert had worked on fortifications for Washington, D.C., and Philadelphia. He had developed river harbors in the upper Midwest and had improved navigation in New York's Long Island Sound. He'd built locks on the Tennessee River.

He was an engineer of a different kind as well. In 1876, not yet fifty, Colonel Robert had published *Robert's Rules of Order*, the standard manual on parliamentary procedure.

In Galveston, Colonel Robert designed jetties, built into the bay to raise the water level. Sand and wreckage were dredged out of the bottom, lowering it. In 1896, when the world's biggest cargo ship steamed through the channel to a 100-gun cannon salute, Galveston's sandbar problem had been solved. The city had a world-class harbor.

City leadership addressed another challenge posed by nature. A lack of fresh water was built into the island's ecosystem. For years, cisterns had caught rainwater, but reliable supply meant carrying water over from the mainland by boat. Sanitation, among other things, suffered. It was hard to imagine a major city so reliant on outside sources for such a vital resource.

By 1895, this problem, too, was solved. A modern system was in place. Water flowed continuously from a site eighteen miles away on the mainland through a pipe, under the water of Galveston Bay, and into a pumping station on the island.

The DWC thus became a kind of permanent partner to elected government in Galveston. Men like John Sealy, William Lewis Moody, Leon Blum, and Harris Kempner adopted the mode of corporate conglomeration that was coming to characterize the nation's business as a whole. They shrewdly harmonized their business among the wharves, the railroads, and the fleets of sail and steamships. They financed all of those concerns via their own banks. The 1890s saw further civic triumphs for Galveston over nature: another railroad bridge, as well as a wooden wagon bridge to free carting from reliance on the railroad. People now passed easily from the booming island city to less sophisticated mainland haunts. In overcoming natural adversity, Galveston had bested all of its neighbors.

But nature gave Galveston another problem: storms. Some might have seen this one as the biggest problem of all. In 1887, a letter to the *Galveston Daily News* put it this way:

> *There are today untold millions of Northern capital looking southward for investment, of which Galveston would receive her legitimate proportion if we could offer a reasonable argument that the island will not one day be washed away.*

"Washed away": stark words. It was glaringly obvious that the entire city lay at sea level—directly in the path of wild gulf storms. The rail, the bridges, the port, the water and electricity, the ships, the entire cotton trade—all of it seemed to offer a bright future. But gaining Wall Street money depended on convincing investors that Galveston had nothing to fear from the weather.

So Galvestonians came up with a solution to that problem too. They took any suggestion that the city might one day wash away as a slur, an insult to civic pride. Regarding the many storms that did, in fact, periodically knock their city and the whole Gulf Coast around, they took an attitude of defiance, even amusement.

Yes, those storms caused problems. But they had led to very few deaths so far, and there had been no irreversible loss of property. The jovial tone was summed up in a Galveston newspaper story after a wild 1886 storm: "The town got a pretty thorough drenching and a good shaking up," the paper reported, "but is doing business at the old stand, as gay as a lark and as spruce as a grass widow." Later that year, the paper reported on another "squall" that "frightened a good many people of Galveston at first and subsequently entertained them."

Storms were real, everybody knew that. But at the same time, they were nothing to worry about. In that "squall" of 1886, Indianola, Texas, on the nearby mainland, was demolished—yet again. Indianola's citizens had taken enough. They finally abandoned Indianola as a city.

Far from seeing in that town's fate a lesson for themselves, Galvestonians took comfort in it. What happened to Indianola proved something Galvestonians had long believed. Because Galveston was an island, with a bay between it and the mainland, the city was buffered from the worst havoc of storms. The bay would always absorb the shock of a hurricane. That idea got new support when a renowned national weather authority, Matthew F. Maury, stated that laws of physics were such that storm waves simply could not hit Galveston Island with direct force. Galveston was specially blessed. Thanks to the shallowness of the bay waters and the sandbars that ran parallel to it, the city existed in a "cove of safety," as people called it.

What the letter writer of 1887 had said was quite specific. Investment would not come, he noted, without "a reasonable argument" that the island was safe. That argument was now in place.

Galveston's solutions to other challenges posed by nature involved an amazing combination of political and financial clout, civic and business vision, and technological enterprise. The solution to this final problem, however—hurricanes, the problem with the deadliest potential—was sheer denial.

CHAPTER 4

STORM WATCHER

UNLIKE MANY OF HIS FELLOW GALVESTONIANS, ISAAC CLINE wasn't given to bravado. His attitude toward weather, and the damage it could cause, was anything but cavalier. On the roof of the E. S. Levy Building, where Galveston's weather instruments were installed, Cline was watching the sky over the gulf during that first week of September 1900.

The sky was big, blue, and windless. Cline observed it and read the instruments with his usual keen interest. He did this many times a day, every day. Little had ever escaped his attention.

And in that early part of the first week of September, Isaac Cline felt no alarm. He was Galveston's chief meteorologist, head of the entire Texas section of the U.S. Weather Bureau. His weather station was on the top floor of the Levy Building. It was up to Isaac Cline to know—and sometimes to sense—developing trouble, to report it, and to manage it.

But right now, he had nothing out of the ordinary to report or consider. Like those captains at sea making early sightings of the developing storm, Isaac Cline saw in his immediate surroundings nothing that weathermen hadn't seen before, nothing to suggest imminent disaster.

Isaac Cline wasn't just any weatherman. He was one of the nation's top authorities on storms. Nor was he fearful of making bold predictions. Fastidious and exacting, yes: every good weather observer and forecaster was that. But more than once since coming to Galveston, Cline had risked the wrath of his bosses at the Weather Bureau in Washington, D.C., by forecasting disaster on his own authority.

So Cline, at thirty-nine, was a hero in Texas. His forecasting from the weather station at the Levy Building had saved a lot of lives. People often mocked the Weather Bureau: its reliability in making forecasts struck many as sketchy. But people in Galveston, and throughout Texas, placed their trust in Isaac Cline.

Cline, in turn, placed his faith in his own exacting standards of professional conduct. And he trusted the deductive talents that inspired him to save lives with such bold and accurate predictions. He trusted, most of all, in the laws revealed by modern science, especially meteorology.

He was highly trained in that science, and he'd already contributed to it. His professional climb had tracked with the rise of the science of meteorology itself, and with the rise of the U.S. Weather Bureau, which handled that science for the booming young nation. By advancing the science of meteorology, Cline believed, he and others could work wonders in improving life for people around the world. The United States, he believed, would lead the way.

He didn't begin life in Galveston. Isaac Cline was yet another outsider making the island city and the biggest state what they were. His weather career had begun in July 1882, when he was twenty years old and stepped off the train at the Baltimore and Potomac Railroad Station at Sixth and B Streets NW, in Washington, D.C. The first thing he saw was the marker where President Garfield had been fatally shot the year before.

For that and other reasons, the young man found arriving in the capital at once daunting and thrilling. Cline's ambitions for serving his big, booming nation were always high. Washington was, to him, the most important place in the world, and from here he would take up the task of bettering the world through reason and modern science.

So he was delighted to have been accepted into the meteorology training program of the U.S. Army Signal Corps at Fort Myer in Arlington, Virginia. To begin his training, he was to report promptly to the Office of the Chief Signal Officer of the U.S. Army, for the Weather Service was then under the aegis of the Army Signal Corps.

It made sense. The connection between weather forecasting and military operations is natural on many levels, and the main idea behind the Weather Service was to ensure that during wartime, a handpicked set of young scientists, fully trained in meteorology, could be quickly attached to the Signal Corps to make weather predictions. The Signal Corps itself, formed during the Civil War, was already a leader in communications technology. By now it had the major responsibility for maintaining and operating the nation's thousands of miles of telegraph lines. Telegraphy, with its amazing ability to transmit detailed information over great distances with lightning speed, already formed a critical component of weather reporting.

With the U.S. military now enjoying a lock on the massive, national-scale organizing power that weathermen needed, "military discipline would probably secure the greatest promptness, regularity, and accuracy in the required [weather] observations," as the U.S. Congress put it when it formed the Weather Service and placed it under the authority of the Signal Corps. Weather forecasting, military prowess, and great nationhood went together.

On November 1, 1870, the Weather Service therefore went into operation, producing its first meteorological reports. They were based on observations made by sergeants at twenty-four stations around the country and transmitted by telegraph to a central office in Washington, D.C. This national report was, in the terminology of the day, "systematized and synchronous." It gave, that is, a single, integrated snapshot of the weather across the whole continent, based on conditions observed locally.

Eleven years later, when Isaac Cline arrived in Washington, people took for granted what had once been a miracle: daily Weather Service reports. People had even started complaining about those reports.

Cline didn't know it then, but the Weather Service was hitting some hard times. He would play a role in repairing its reputation.

The prospect of his new career could not have been more exciting. Yet he'd never been in a big city before. That first night in Washington, Cline was afraid to stray out of sight of his hotel. He was just out of school, having received his B.A. from Hiwassee College, only five miles from where he grew up

on a farm in Monroe County, Tennessee. He'd stood out at Hiwassee, and some of his professors had sponsored his ambition
to become a scientist. To pay his way, he worked in the college
library and chopped wood for the school's fireplaces. He went
home on weekends to work on the farm and attend church; on
school vacations, too, he worked on the farm. Until he became
a scientist whose career took him around the country, Isaac
Cline remained a Tennessee country boy.

While the Civil War was hard on his part of the state (the
Clines and many others there were pro-Union), with the end of
the war, Isaac's father was able to build up his farming properties. His ambition was to get his sons educated and to give each
of his daughters a farm upon her marriage.

The father's plan succeeded. When, after graduating from
Hiwassee, Isaac boarded the train in Sweetwater, Tennessee,
for Washington and a new life, his father gave his blessing between sobs.

At the Fort Myer training school in Arlington, Isaac Cline
wore an army uniform, slept in barracks, and drilled in squads
with the other young meteorology recruits. He served guard
duty. He learned infantry and cavalry tactics, horsemanship,
and signaling with flags, torches, and balloons. He studied the
mechanics of both the telegraph and the telephone.

But mostly Cline and the others learned how to take, record,
and communicate meteorological readings. The instruments
they studied became the tools of Isaac Cline's science. With
only slight refinements, these were the same tools he would be
using in Galveston in 1900.

Cline took up the study of these instruments with great
application. Here was the technology that, combined with methodical, precise, and objective observation, would allow hu-

mankind to take control of the biggest forces in nature. With these few tools, a single man could predict floods, tornadoes, and hurricanes and help thousands of others respond to them in advance. Nature's terrors would succumb to the superior intelligence of the human race.

Nineteenth-century weather instruments, though not infallible in themselves, really are amazing. The complete weather station that Isaac Cline learned to install and use had a number of integrated elements.

They began with the simple weathervane, which shows wind direction and allows calculation of that direction to exact degrees. A rain gauge, also fairly simple, catches precipitation and measures it in a tube.

More complex was the sunshine recorder. A glass sphere focused the sun's rays on cards mounted at the sphere's back. The sun burned the card, allowing weathermen to calculate the number of hours of bright sunshine in a given day.

But those instruments measure only what is happening now and what has happened recently. The big meters—anemometer, barometer, and hygrometer—involve not just measuring but forecasting, and especially forecasting disasters. While the science of forecasting was becoming, in Cline's day, a modern and objective one, much of the technology on which it depended was ancient.

Of the big three, the anemometer used the oldest technology. Four fine, metal, hemispherical cups, their bowls set vertically against the wind, caught air flow. Because each cup was fixed to one of the four posts of a thin, square metal cross, lying horizontally, and because the cross's crux was fixed to a vertical pole, when wind pushed the cups, they made the whole cross rotate. It made revolutions around the pole.

In Cline's day, the pole was connected to a sensor with a dial read-out display. The number of revolutions the cross made per minute—clocked by the sensor, transferred by the turnings of the wheels, and displayed on the dial—indicated a proportion of the wind's speed in miles per hour.

Rotating cups, wheels, and a dial: the anemometer was fully mechanical, with no reliance on electricity. And while other competing anemometer designs existed, involving liquids and tubes, the four-cup design became standard in American meteorology in the nineteenth century, remaining remarkably stable.

In 1846, an Irish meteorologist named John Thomas Romney Robinson upgraded the technology. But before that, the biggest development in clocking wind speeds had been made in 1485—by Leonardo da Vinci. The anemometer was already a durable meteorology classic when Isaac Cline began studying.

The second member of the forecasting big three, which Isaac Cline studied with such interest under the Signal Corps, was the hygrometer, which measures relative humidity. Like the anemometer, it's been around ever since a not-very-accurate means of measuring relative humidity was built by—once again—Leonardo da Vinci.

By Cline's day, a basic hygrometer measured the degree of moisture in the air by using two glass bulbs, each at one end of a glass tube. The tube passed through the top of a wooden post and bent downward on both sides of the post, farther down one side than the other. Thus one of the bulbs was lower than the other. In that lower bulb sat a thermometer, dipped in ether, a gas that had condensed in the bulb into a liquid.

The other, higher bulb contained ether too, but here the gas remained in its vapor form. That bulb was covered in a light fabric.

When condensed ether was poured over the fabric covering the higher bulb, the bulb cooled, and the vaporized ether within condensed, lowering vapor pressure in the bulb. That lowering of pressure caused the liquid ether in the lower bulb to begin evaporating into the space provided. So the lower bulb's temperature fell as well.

Moisture—known as a "dew"—therefore formed on the outside of the lower bulb. As it did, the temperature indicated by the thermometer in that bulb was read and noted. That reading is called the dew-point temperature. Simply comparing the dew-point temperature to the air temperature outside the bulbs—as measured by a common weather thermometer, conveniently mounted on the hygrometer's wooden post—gives the relative humidity. It's a ratio of dew-point temperature to air temperature. The closer dew-point temperature gets to air temperature, the higher the relative humidity.

As a student of humidity, Isaac Cline read tables (sometimes built into the hygrometer for quick reference) showing the exact humidity ratios. But experienced forecasters know the rough ratios by heart.

We concern ourselves with humidity mainly on hot days. When there's lots of moisture in the air, it can't accept much more moisture, and that means warmth has a harder time leaving our bodies via perspiration. With a temperature of 95 degrees Fahrenheit and a dew point of 90 degrees Fahrenheit, you'll get a relative humidity of nearly 86 percent—quite uncomfortable. When air temperature and dew point are identical, humidity is said to be 100 percent. We really don't like that.

There were other kinds of hygrometers as well, developed during Isaac Cline's early career as a meteorologist. One, called a psychrometer, compared a wet thermometer bulb, cooled by evaporation, with a dry thermometer bulb.

And in 1892, a German scientist—he had the unfortunate name of Richard Assmann—built what was known as an aspiration psychrometer for even more minute accuracy. It used two matched thermometers, protected from radiation interference by a thermal shield, and a drying fan driven by a motor. By 1900, when Isaac Cline was working in the weather station in the Levy Building in Galveston, hygrometer science was at its apex.

But perhaps the most important element in weather forecasting is the barometer.

The role of barometric pressure—air pressure—is counterintuitive. We can directly feel the phenomena measured by anemometers and hygrometers—wind speeds and relative humidity: wind knocks us around and humidity makes us sticky. But the sensations caused by air pressure work differently from the way we might expect.

That difference has to do with the very nature of air. Usually we don't think much about air. While we know it gives us oxygen, breathing is largely unconscious. We notice air when it's very still or very windy. And we notice air when it stinks.

Otherwise we generally ignore the air. We imagine it as nothing but a weightless emptiness.

But air does have weight. That weight exerts pressure on the Earth's surface, as well as on everything on the Earth: human skin and inanimate objects. We refer to the pressure of that weight as "atmospheric pressure," and we measure it with a barometer.

When more and bigger molecules gather, air's weight increases, and the atmosphere bears down snugly on all surfaces. We call that effect, not surprisingly, "high pressure." The strange thing, though, is that high pressure—all that heavy weight of air—makes us feel freer, more energetic. It makes the air feel not heavier but lighter.

That's because where pressure is high, relative humidity is suppressed. Warmth can't lift as easily from the surface of any object—including from the Earth's surface. Warm air currents are held at bay, moisture is blocked, winds remain stable. Rain, lightning, and thunder are discouraged. High pressure usually means nice weather.

By the same token, when we complain that the air feels heavy—on those days of sluggishness, when we feel as if we're struggling through a swamp—heaviness is not really what we're feeling. Just the opposite. On those days, the air has less weight, lower pressure.

The result, usually, is just some unpleasantness. That's because lighter and fewer molecules in the atmosphere cause atmospheric "lifting." Heat and moisture lift upward from all surfaces. The humidity gets bad.

But when barometric pressure falls low enough, winds may be expected to rise, clouds form, and rain, thunder, and lightning follow. With very low air pressure, things aren't just unpleasant. They're dangerous, sometimes deadly.

Barometers for measuring pressure had been part of experimentation in natural science since the 1640s, well before modern weather forecasting. For a long time, it just seemed interesting, and possibly useful, to know that atmospheric pressure exists at all. Or to see that it can do work—like pushing mercury upward in a column.

But soon people began to apply the science. They used pressure readings not only to note the existing weather but to predict future changes in weather. One scientist graduated the scale so the pressure could be measured in exact increments. Another realized that instead of pushing mercury upward, the scale could be turned into a circle to form a dial; that enabled far subtler readings.

Yet another change came with the portable barometer. Using no liquid, and therefore easier to transport on ships, the portable barometer took the form of a small vacuum-sealed metal box, made of beryllium and copper. Atmospheric pressure made the box expand and contract, thus moving a needle on its face. A barometer like that could be carried in a pocket by a ship's captain. He could watch the pressure fall and know that he was sailing into a storm.

Just before Isaac Cline began studying, the widely traveled Vice-Admiral Robert FitzRoy, of the British Royal Navy, formalized a new system for detailed weather prediction based on barometric readings. FitzRoy had served as captain on HMS *Beagle*, Charles Darwin's exploration ship, and also as governor of New Zealand. His idea was to go beyond just noting existing and future weather conditions. He found ways of communicating conditions from ship to ship. That aided safety at sea.

By the mid-nineteenth century, a large barometer of FitzRoy's design was set up on big stone housings at every British port. Captains and crews could see what they were about to get into. In 1859, a storm at sea caused so many deaths that FitzRoy began working up a system of charts that would allow for what he called, for the first time anywhere, "forecasting the weather."

Using the telegraph, which had recently become feasible for this kind of work, FitzRoy established an interconnected group of weather stations that made daily reports to a central station. Soon he was directing a system of storm-warning cones, hoisted at major ports, in response to information coming in from all stations. Boats could be ordered to stay at port under certain conditions. Using terms like "gale warning" and "small craft warning," the modern weather bureau, familiar for decades to come, had been born.

And all of that was based on using a barometer for measuring rising and falling air pressure.

The military discipline of the training school at Fort Myer, Isaac Cline felt, fit neatly with the unflagging regularity, neatness, and precision that he began to see as critical to good weather reporting. Along with the use of instruments, the students studied the classic Weather Service and Signal Corps texts: *Loomis's Meteorology, Myer's Manual of Signals, Instructions to Observers, Pope's Telegraphy,* and *The Handbook for the Signal Corps.* And they took strict exams on both theory and practice.

As graduation neared, what Cline and all of his fellow trainees wanted most was a job as assistant weather observer at a Weather Service post. Only sixteen men passing the Fort Myer course with the highest grades would get such an assignment.

Cline passed sixteenth.

He'd barely made the cut, but he was an observer. Cline was immediately assigned to the weather station at Little Rock, Arkansas. The Rocky Mountain locust had been ravaging the countryside. Cline's assignment—along with taking and reporting weather readings—was to study the influence of weather conditions on the development and migration of that destructive insect.

By the time he arrived in Galveston, Texas, in 1889—only six years after taking up his first assignment in Little Rock—Isaac Cline possessed a remarkable trove of weather expertise. The

Rocky Mountain locust had mysteriously disappeared from the West. Cline never got to know much about it.

But his barely making the cut as an observer certainly didn't hold him back. Maybe it even inspired his commitment to excellence. In his early years as a weatherman, Cline learned how to make extremely accurate observations and remarkably reliable forecasts. When the weather observer at Fort Smith, Arkansas, died unexpectedly of a heart attack, Cline was sent there to take over temporarily. Soon he was in charge of the station at Fort Concho in Texas.

The move to Fort Concho began Isaac Cline's long and fruitful relationship with the nation's biggest state, which was just on the verge of its fabled boom. A spur from the mighty Chisholm Trail brought thousands of head of cattle thundering into market in Fort Concho. The Texas & Pacific Railway had recently made nearby Abilene a stock-shopping point—the classic cowtown.

Also nearby, San Angelo had only recently been a lawless boom town full of gunfighters and saloon poker games. Range wars among various kinds of ranchers and family farmers, old feuds between Rebels and Yanks—all contributed to the wild mood of the Texas frontier.

But in contrast to the rowdy action around him, what Isaac Cline did all day was take readings, three times a day, from the barometer, the thermometer, the anemometer, the weathervane, the rain gauge, and the hygrometer; and combine those readings with eyeball observations of the type and movement of clouds and the amount of sun and rain. Then he would code those observations and telegraph them to Washington.

He had to be sure to be take the observations at exact times in order to allow Washington to coordinate them with readings taken at those same exact times at all the stations in the country. Perfect coordination contributed to a simultaneous

observation of the continent at 7:35 A.M., 4:35 P.M., and 11:00
P.M. Washington time.

But Cline also took three other observations at local time:
7:00 A.M., 2:00 P.M., and 9:00 P.M. He took a seventh observation at noon. If anything seemed worrisome at noon in com-
parison to earlier readings, he was to telegraph his concern to
Washington.

Then, exactly at sunset, he took another observation, describing the appearance of the western sky, the wind direction,
cloudiness, barometer, thermometer, and hydrometer readings,
and rainfall since the preceding report. That was telegraphed
with the late-night report.

The office in Washington put together all this punctilious
observation and simultaneous reporting from a multitude of
stations in big sections of the country and sent forecasts to
thousands of rural and urban post offices around the country. Above each post office, huge signal flags—up to six by
eight feet—relayed the forecast to farmers, businesspeople,
and anyone else who needed to know. Reports also went to
newspapers.

At the weather stations, regulations were strict, thanks in
part to the military discipline under which the entire service
operated. And Isaac Cline was a full supporter of that strict-
ness. He required complete adherence to the rules—from himself and, as he moved up the Weather Service ladder, from his
assistants. Clerks were to keep desks, drawers, and file cases
neat and clean. No paperwork was to be strewn about: all
papers were to be returned to their files immediately after use
and at the close of each day's work. Every Saturday, a property
officer policed and set up all rooms, halls, stairways, closets,
and cellars.

There was to be no taking medicines at the watercoolers.
No talking outside about what goes on at the office. No visi-

tors, except on business, and visitors on business must transact it, then leave. No casual conversation. No writing private letters. No reading the paper. Business conversation only—carried on in a low tone of voice.

So it's funny to think of Isaac Cline, with his fussy dedication to precision, responsibility, and routine, exemplifying anything of the improvisational spirit of young Texas. He'd quit smoking after his first few cigarettes. He avoided playing poker, first playing only with no ante and then breaking the habit altogether. He disapproved of the deleterious effects of alcoholic beverages. An abstemious weather forecaster doesn't seem the ideal exemplar of growth in Texas.

And yet Cline was an army officer, a representative in Texas of federal authority. As such, his role was special. And Signal Corps officers were cocky. They didn't wear uniforms. Even when under direct military orders, they were empowered to act as mavericks (originally a Texan term for wayward cattle). They carried out operations as they saw fit and liked to think of themselves not as normal army but as specialists, like the Army Corps of Engineers.

So Cline had his own kind of genteel machismo. It was based on his sense of himself as a gentleman, a federal agent, a scientist, and a doctor. And while a professional weatherman might not have made much of an impression on a pistol-happy drunk reeling out of a Fort Concho saloon, when Cline arrived in Texas, both the Weather Service and its parent, the Signal Corps, were in fact playing important roles in the key conflicts of the American West.

Indeed, when a Weather Service associate of Cline's, Charles F. von Herrmann, became attached to the U.S. Army in the

Southwest, he amazed the regular troops with the success of his innovations. This was during the effort to defeat Geronimo, the Apache leader who continued to elude the U.S. Army. Into the hunt for Geronimo, the Weather Service introduced heliography, a signaling system using no wires but only sunlight on a mirror. Flashing in Morse code via quick pivots of the mirror, a heliographer can send messages instantly, across great distances, where no other means exist.

Isaac Cline's colleague von Herrmann thought heliography might serve as an ideal way of fighting Indians in the deserts and plains of New Mexico and West Texas. At first the system just seemed silly to many of the U.S. soldiers bent on tracking down Geronimo's men. Heliography looked like a low-tech gimmick with no useful application.

Von Herrmann, undeterred, commanded army squads in setting up a complex system of heliographs in the theater of war. The New Mexican and Texan deserts possessed not only long-range, unobstructed views but also a degree of dryness in the air that made tiny details startlingly clear. Von Herrmann sited the mirrors on the bluffs and buttes above the flat plain. As he moved from mirror to mirror around the countryside, he became a bizarre sight. To protect his peeling red face from the New Mexico sun, he was wearing a wire frame secured to his waist and holding a white canvas tent over his head.

After much time-consuming experimentation—during which he had to suffer the skeptical mutterings of his troops—von Herrmann staged a triumph. The men stood amazed as he sent a comprehensible message ninety miles instantaneously. By setting up relays, the mirrors could soon send similar messages over 200 miles.

Some U.S. generals later claimed that the Indians under Geronimo were so amazed by the flashes of light that they were awed into surrender. But Geronimo was well acquainted with signaling over long distances, and he'd seen light and mirrors before. The generals' claim seems more than unlikely.

More realistically, the troops themselves noticed that the Indians stayed away from areas where they knew heliographs had been stationed. Geronimo and his men knew exactly what the mirrors did: talking with sunlight, as Geronimo himself later put it, making it impossible for him to elude capture.

Not only aiding communication between U.S. troops, but also tracking and mapping Indian troop movements over a huge expanse, heliographs helped the United States close in on Geronimo. The modern science of meteorology gave the United States a tactical advantage that it had been lacking in the Apache wars.

Von Herrmann's next assignment was to pack up Isaac Cline's Fort Concho weather station and safely transport all the delicate instruments to Cline's new assignment. This was in that once wild cowtown, rapidly becoming a real city: Abilene itself. There Cline launched what the Weather Service was now calling a First Order Weather Observation Station. That meant the station would employ all of the latest technology in use by the service at that time. The investment was a sign of federal confidence in the importance of Texas.

Cline's First Order station in Abilene made use of self-registering instruments. Too expensive for smaller and less important stations, self-registration added a tricky and impressive feature to the classic tools of weather observation. In less expensive arrays, a weatherman has to write down all of the readings, taken at every interval, and keep them in tables. Downsides are the time involved, the tedium, and the natural potential for human error.

But by connecting a nib on a wire to the instrument's dial or movement, a machine could be forced, by the very laws of physics that were creating the reading itself—evaporation of ether through a hygrometer tube, water rising in a rain gauge, pressure expansion and contraction in a barometer—to inscribe a mark on paper. That paper, thanks to tiny wheels moved by similar action, might unwind from a spool or slide vertically as the marks were made. With clockwork, the marking of the pen and the movement of the paper could be related with exact precision to the passing of time. And since, by the late nineteenth century, such mechanics could also be enhanced by motors and electricity, there was potential to bring down the costs of operating and maintaining these delicate yet labor-saving self-registering instruments.

In Abilene, Isaac Cline was not only able to make minute observations automatically but also to study the old rainfall and temperature records of the area. On that basis, he began to publish studies that became instant classics of modern scientific weather observation. Their titles don't sound exactly thrilling: "Precipitation and Its Sources in the Southern Slopes," "Summer Hot Winds on the Great Plains," and so on. But they show his growing interest in becoming not merely a fine weatherman but also a major contributor to the science itself.

So by the time he came to Galveston in 1889, Isaac Cline had operated the most sophisticated weather forecasting technology of his day. He was a published expert in all kinds of weather phenomena. And he was developing a special interest in hurricanes.

Not all of the experience Cline brought to Galveston was meteorological. Back in Little Rock, finding himself with time on his hands, he'd taken up the study of medicine. Somewhat amazingly, he was now in fact, an M.D., having attended the University of Arkansas Medical School, only three blocks from the Little Rock bank building that housed his first weather station.

Those two vocations, meteorology and medicine, complemented one another ideally, Cline felt. For one thing, he said, each gave his mind rest from the other. That's an indication of Isaac Cline's approach to recreation.

He was writing his weather studies on his own time. He called that recreation too.

But the main thing Cline liked about combining medicine and meteorology: he wanted to understand connections between weather and human illness. He wanted to use meteorology not only to help the army, and not only to provide accurate, useful information to ordinary American citizens. He most wanted, as he forthrightly thought of it, to aid mankind in its progress toward a better world. That better world was beginning to seem inevitable as Cline came of age, and to seem inevitably American.

As if he weren't busy enough, Cline got also into the newspaper business in Abilene. A man named Sayles, a lawyer and former Confederate general, had his law offices near Cline's First Order Weather Observation Station. Sayles owned the *Abilene Daily and Weekly Reporter*, and Cline (for recreation) wrote articles on local weather observations. From Sayles's rewriting of his pieces, Cline learned much about clear written expression.

Soon Sayles invited Cline to take over editing and publishing the paper. Cline accepted, "as something to do in my recreation time," he later explained. He set up an entire new print shop for putting out the news. The shop began making money printing other jobs too.

Meanwhile, cattle ranchers and sheep ranchers were duking it out on the range and in town. The former wanted free trade, the latter a tariff. Both supported Cline's paper, and both wanted to control its editorial policy. These were rough men. But for all of his seemingly non-Texan abstemiousness, Cline remained unafraid.

He kept a loaded .45-caliber Winchester rifle, with a hair trigger, leaning against the wall near his desk in the newspaper office. In the end, he never had to use it for anything but hunting.

One of Cline's editorial positions at the Abilene paper mixed his publishing efforts with his real job. He favored efforts to establish a Texas Weather Service, overseen by the state. Cline's boss at the U.S. Army Signal Corps, General Adolphus W. Greely, agreed. A dedicated service for that huge region would have many advantages.

But a law to create the state Texas Service failed to pass the Texas legislature. When Cline reported that failure to Greely, the general had a new idea.

Greely told Cline he was creating a division of the U.S. Weather Service, a branch solely dedicated to Texas. This new division would have its headquarters in Galveston, the most important city in the state. Greely wanted Cline to be in charge.

Cline was twenty-seven. The former Tennessee country boy, fired as a youth with the ambition to serve his rising nation and the betterment of the world, handpicked to fulfill that ambition by studying meteorology, would spend eleven years in Galveston. There he would refine those abilities while serving the city, the state, and the nation. He had good reason to be supremely confident in the scope and quality of his many abilities.

In October 1890, soon after Isaac Cline's move to Galveston, the U.S. Congress passed a sweeping law that transferred the entire Weather Service, with all of its employees, stations, and equipment, out of the army altogether. The service became part of the Department of Agriculture.

This was a benchmark moment both in the history of the service and in Cline's own career. For if the Galveston assignment was a culmination in the progress of his rapid journey up the ranks of the Weather Service, the service itself, even younger than Cline, was making big strides too. With the transfer to the Department of Agriculture, the Weather Service was well on its way to becoming one of the most important agencies of the civilian federal government. Isaac Cline's position in Galveston, overseeing meteorology for the entire state of Texas, would make him one of the most important federal officers in the country.

When the transfer occurred, General Greely was still Isaac Cline's boss, and Greely was widely considered to have saved the service. Before the general's time, the service had seen some rough years. During Cline's first years in the organization, the service had become highly controversial, constantly investigated for fraud, scandal, and corruption. A Captain Henry W. Howgate, for example, the disbursing officer of the Signal Corps, had embezzled over $200,000 by submitting fake vouchers from contractors.

People believed members of the Weather Service had aided and abetted that rip-off. A War Department investigation revealed that, at the very least, the service had no real financial oversight. The Chicago Board of Trade petitioned Congress to reform the Weather Service. Highly placed people said the training school at Fort Myer should be shut down.

Meanwhile, army brass were getting sick of all of the Signal Corps officers' cockiness—especially that of the weathermen. When, for example, von Herrmann, the heliograph expert who helped find Geronimo, first reported for that duty, his commanding officer began giving him the usual orders regarding barracks, mess times, and so forth, and von Herrmann interrupted to set the officer straight. Weathermen were not to be ordered about, von Herrmann reminded his C.O. Herrmann would sleep and eat where he pleased—now would the officer get on with it and assign troops to work on the heliograph?

That's not the kind of attitude regular army liked to see. Furthermore, the War Department was waking up to the fact that the Weather Service had essentially taken over the Signal Corps. The whole corps was really nothing but weathermen now. If the army ever needed actual signaling, nobody would be available.

The accuracy of the forecasting, meanwhile, had become slipshod and unreliable—at least in the collective mind of a citizenry dependent on it. When the Weather Service began in 1870, simultaneous reporting and forecasting for the whole nation had been something of a miracle of modern science. And in fact the miracle only grew. By 1885, predictions were being made for the next thirty-two hours. In 1886, forecasts were made for states, and even parts of states, instead of big districts. Soon forecasts were extended for forty-eight hours.

Yet now people expected these forecasts, and they therefore often found them disappointing. For of course the reports weren't always right. Sometimes dramatic events occurred that hadn't been forecast, and sometimes, as in the Northeast blizzard of 1888, those events killed people—400 people in that case. Meanwhile, other events that had been forecast failed to materialize.

There was a sense that the inaccuracy might be resulting from a failing of the strict discipline and observational precision that Isaac Cline himself loved so much. So when General Greely took over, he was determined to address the discipline issue.

The general sent inspectors far and wide. Inspections of weather offices resulted in one hundred firings of weathermen. Isaac liked regulations, but many of his colleagues apparently didn't. Some of them entertained, putting it politely, non-business visitors in the observation office. One weather officer in New England was using the office to shoot photos of nude women.

Another neat trick was to invent a full week's worth of fake observations. Just hand them to a telegrapher in a batch. The telegrapher feeds them to Washington day by day at the appropriate times. You get in a week of fishing.

One observer pawned all his weather instruments to raise cash for gambling. At least that one kept up his reports: he went down to the pawnshop every day to take readings.

Soon Greely gave up. Under current circumstances, he told Congress, the weather reporting situation was not to be reformed. President Benjamin Harrison recommended transferring the Weather Service to the Department of Agriculture, and Congress promptly did so. The organization was now called the United States Weather Bureau.

Isaac Cline had a passion for discipline and regulation, a need to be ceaselessly working, and a deep desire to use his expertise in aid of greater American and human projects. Those qualities had long made him stand out in what was a troubled Weather Service. When Greely sent Cline to Galveston and expanded the Galveston office into a headquarters for a dedicated Texas department, the general was rewarding the young man for excellence in his profession.

Greely was also relying on Cline to help improve the entire Weather Bureau, for during inspections, the Galveston office had turned out to be among the worst. Located then in a building that also housed the police station and the courthouse, the office was more like a bad kennel, the inspector reported, than an office.

In order to read the barometers, the inspector had to clean them. He also noted that the citizens of Galveston ignored reports coming out of their own weather station. Instead, they checked the St. Louis and New Orleans papers.

Isaac Cline was there to change all that—and to expand the station into a major operation overseeing all of Texas. Along with daily readings, the office had to produce the monthly Texas Weather Bulletin in conjunction with the Galveston Cotton Exchange. Isaac Cline had thus become section director of the Texas Section of the U.S. Weather Bureau, executive director of the Galveston Weather Station, and executive director of the Cotton Region Services for Texas. To make all that work, he would need assistants. He got busy.

He also settled with great pleasure into bustling, growing Galveston. Back in Abilene, he'd married Cora May Bellew, whom he'd met in church. She'd been playing the organ at the time. The church connection, along with Cline's overall punctiliousness, might have made the pair seem the ultimately straitlaced couple, but when their first child—a daughter—was born, they'd been married less than nine months.

In Galveston, they had a second daughter, and then a third. They moved into a large, sturdy, well-built house about a block from the gulf beach.

Family and work overlapped. Cline hired, as his chief assistant in Galveston, his younger brother Joseph. Joseph Cline had knocked about a bit after following in his brother's footsteps and graduating from Hiwassee; he'd worked as a schoolteacher and traveling salesman and served a stint on the railroad before Isaac hired him.

But Joseph became a good weatherman. He would play an important part in the wild events of September 1900.

Isaac Cline stayed busy, to say the least. Father of three, buried in the herculean task of turning the disastrous Galveston weather office into a national flagship for a new administration, he nevertheless felt the usual need for what he called recreation. This time it took the form of teaching medical climatology at the University of Texas Medical School in Galveston. Meanwhile, he matriculated at the AddRan Male and Female College (named after the Clark brothers, Addison and Randolph). There he studied philosophy, sociology, and English literature. Soon he had his Ph.D. He credited his improved understanding of literature with aiding him in his weather observations.

But he found he had free time on Sundays. So he taught Sunday school.

And very quickly, the Galveston headquarters became everything Cline and General Greely had believed it could be. Isaac and Joseph, working together, turned the station into a ship-shape operation handling more data on a daily basis than any other in America. Barometers were plumb, their vacuum seals tight, their dials clean. Daily readings were precise, detailed, and realistic. Resulting forecasts were more reliable. People in town and across the state began noticing. Under Isaac Cline, the Galveston office wasn't merely providing excellent service. It was gaining respect, throughout Texas, for the entire U.S. Weather Bureau.

The office passed its first inspection under the Agriculture Department with flying colors. In his report, the inspector went out of his way to note that nobody in the field was better than Isaac Cline. Joseph too was deemed an admirable meteorologist—although his penmanship needed work, the inspector remarked. The Clines were doing an exemplary job, the inspector went on to say, handling with aplomb more information than any other office and meanwhile raising the profile of the bureau both in town and throughout Texas. Isaac's mission in coming to Galveston had been accomplished.

In 1900, Isaac Cline, now thirty-nine, had long been a well-respected man around Galveston. A nationally recognized weather expert, with a multitude of important publications, running the biggest station in the nation, he'd continued to take his usual recreational interest in the study of local conditions, and he'd made some brilliant moves.

For one thing, Cline came up with a way to give Texas farmers three days' notice of freezing conditions—and he bucked the central office in Washington to do so. In the capital, they'd said such predictions were impossible; anyway, all predictions were supposed to come from Washington, based on local reports. Information was local; forecasts were national. Local offices couldn't make forecasts.

But Cline went ahead with his studies of local freezing on his own time. The resulting predictions of frosts were accurate enough to wow the bureau in Washington. Cline won official permission to issue cold-weather forecasts on his own.

Another instance of Cline's leadership came in the spring of 1900, when Galveston's weatherman saved untold lives by predicting the flooding of the Colorado and the Brazos Rivers. Again, only Washington could legally issue flood warnings, but Cline issued them anyway—and in the case of the Brazos, he called the flood's high point with perfect accuracy, ten days in advance.

That was the biggest flood ever seen on the Brazos. Thanks to Isaac Cline's early warnings, no lives were lost.

Here was the kind of success that close study, maverick intelligence, dedicated science, and bold application could achieve when it came to the weather. That had been Isaac Cline's aspiration; now it was his achievement. Modern man (as people put it then) was approaching a new position. Soon human beings would be able to manage any and all of nature's disastrous threats. Thanks to science, the world was on its way to becoming a safe, happy, and rational place. The twentieth century would be a great one.

So the citizens of Galveston were lucky to have as their resident meteorologist somebody with a proven national reputation for expertise in storms. It's hardly surprising, given Isaac Cline's status as a great weatherman, that civic leaders would seek his opinion regarding the city's vulnerability to hurricanes.

During the Deep Water Committee's dredging project, the great engineer Colonel Robert had proposed building a break-water off the beach, just out in the gulf, to hinder waves and tides. There had also been intermittent talk of building a high seawall on the beach itself to push the waters back in the event of a storm.

The people of Galveston were, of course, used to storms, flooding, and damage, even to a few storm-related casualties every so often. As the city boomed, the civic tradition of jovial disdain for people who fearfully envisioned its washing away overnight continued.

Deep gutters were meanwhile dug in the streets to drain rain and floodwaters out of the city. Near sea level, the gutters were two to three feet deep, shallower as the streets rose to higher ground in the island's interior. They often did a good job of moving water out.

But the question remained. As the letter writer had put it

back in the 1870s, the city needed a reasonable argument to show that it would not one day all wash away.

So in 1891, after a nasty storm yet again caused flooding on the island, the *Galveston Daily News* asked for Dr. Cline's opinion regarding the likelihood of a deadly hurricane's smashing the city. Should the city really go to the expense and effort of building a high seawall along the gulf beach to keep rising gulf waters out of town?

Cline responded with his usual clarity and precision. He based his opinion on a mountain of just the sort of scientific fact and empirical evidence he'd employed to save crops when predicting frosts and to save lives when predicting floods.

It was overwhelmingly clear, he wrote in the *Galveston Daily News*: if a storm ever pushed high gulf surf into the island, the water would merely flow over the city, enter the bay, and then keep rolling onto the mainland plain of Texas. The city itself would barely feel the effect.

Too, the gulf coastline is shallow. It would fragment any incoming surf.

And really big hurricanes didn't strike Texas anyway. "It would be impossible," Isaac Cline wrote, "for any cyclone to create a storm wave which could materially injure the city."

So there it was. According to possibly the best weatherman in the country, the city's own Dr. Cline—who lived near the beach—Galveston was perfectly safe. Citizens who feared otherwise, Cline said, were in the grip of "a severe delusion."

Naturally the city took Dr. Cline's advice. There was no seawall.

CHAPTER 5

THE STORM: FROM CUBA TO TEXAS

ON THE MORNING OF WEDNESDAY, SEPTEMBER 5, THE STEAM-ship *Louisiana* of the Cromwell Shipping Line passed the Port Eads weather station at the bottom of the Mississippi bayou channel, about eighty miles south of the port of New Orleans, nearly 400 miles east of Galveston. The Port Eads station stood near the channel's entrance to the open Gulf of Mexico, and a storm-warning flag had been hoisted there. The wind was blowing hard out of the east-northeast.

But the ship's captain, T. P. Halsey, was a veteran seaman. There had been no reports of any serious weather moving into the gulf from the south. Anyway, a storm was something he and his ship could handle. Halsey noted the storm warning, and his steamship crossed the bar, as sailors put it, and entered the Gulf of Mexico.

With a full crew, heavy cargo, and thirty passengers on board, Captain Halsey's destination was New York City, and so the *Louisiana* was heading southeastward, across the gulf toward the Florida Keys. Keeping the warning flag in mind, Halsey did order the decks stripped and the hatches battened down in preparation for chop. The *Louisiana* went steaming down the gulf for the Keys.

That night was rough indeed, and all day Thursday the ship rose, fell, and rolled, fighting the rising seas, beaten by winds that had started to howl. Thursday night was even worse.

By Friday morning, most of the passengers were too sick to be frightened. They rolled miserably about in their berths, their cabins locked behind storm doors.

Captain Halsey wasn't scared. He'd been through eight hurricanes at sea. He ordered the crew to hold course against the growing violence of the wind and the bucking waves.

On Friday morning, the wind became gale force; it had turned around now, and it came from the north-northeast, with what sailors call a "backing wind"—movement high up at the steamship's smokestacks—turning hard from east to north.

With wind direction now turning in a circular pattern that was unhappily familiar to Captain Halsey, the rain-battered ship continued on its sickening roll.

Then there were the barometer readings. Entering the gulf on Wednesday, Halsey had taken a reading of 29.87 inches: quite normal. Thursday night, the pressure had started falling; now, early Friday morning, the reading was 29.60.

Each tenth of an inch represents a drastic difference in barometric pressure. Halsey's barometer had dropped nearly three. The greater the speed of the drop, the greater the violence of the weather. This three-point drop had occurred overnight. Readings below what Halsey was getting now were considered anomalous in 1900, signs of great disturbance, and yet his barometer kept falling.

Where they were heading, the lack of pressure must be oth-erworldly. Chaos must rule.

So Halsey now knew exactly what was happening. He'd steamed directly into the middle of something that he was prob-ably the first U.S. citizen to call by its real name: a hurricane.

Here at sea, hundreds of miles southeast of Galveston, the crew and passengers of the *Louisiana* were about to ex-perience, more than a full day in advance, what people in the island city still had no idea lay in store for them.

Captain Halsey could not be rattled. That's in spite of the fact that at 10:00 A.M., while his ship pitched and surfed through the storm, he took another barometer reading and saw that it had dropped to 29.25. That was low pressure of a new order, and it represented a shocking plummet. Only four hours had passed since the previous reading.

At 1:00 A.M., the *Louisiana* was in the midst of the hurri-cane. Halsey assessed the winds shaking his great steamship like a toy boat, making it shudder all over as it strained up and down the huge waves. The wind speed, he thought, must have reached 150 miles per hour.

Worse, there was now no single wind direction. The blast shifted crazily from one direction to another, seeming to come from all points on the compass at once, driving sheets of rain against the craft's steel walls.

Nobody knew how long a ship like this could hold out against such punishment. The bow shot upward against each monstrous wave. The ship hung, then the bow crashed back down. Wave after wave washed across the deck, submerging it.

Then, just as the bow was again diving into an alley between two mountains of water, a wave slammed the ship from behind. The entire steamship went underwater, all at once.

But the ship emerged. And it kept going. Another huge wave, slamming the ship on one side, poured ocean into the ventilators, flooding the engine room. Still, Captain Halsey kept steaming. There was nothing else to do now.

For three hours, Halsey's crew kept fighting the hurricane. Anyone glancing at the barometer that afternoon would have looked twice. It reached 28.75. In 1900, many meteorologists would have questioned the veracity of such a reading. Moving through a chaotic realm of elemental violence that most people will never see, Halsey just kept going.

When he brought the *Louisiana* into port at Key West, the storm had finally begun to move on. The cargo had shifted, putting it mildly; Halsey's ship was in need of attention. But all hands and passengers were fine. The *Louisiana* soon steamed on to New York.

Days later, pressed by eager New York reporters, Captain Halsey wouldn't admit that the storm was the worst he'd ever seen. By then everybody knew that it was. For by then the storm was world famous.

Captain Halsey's voyage into the storm began on Wednesday, September 5. The same day, the *Galveston Daily News* ran a tiny, twenty-seven-word squib in its weather section.

Buried in a flurry of small print, unorganized headings, and advertising typical of the crammed broadsheets of the day, this notice was datelined "Washington, D.C.," the previous day, with Isaac Cline's Galveston station cited as the "observer."

Washington's notice in the Galveston paper advised readers of what it called a tropical disturbance in Cuba. The disturbance was moving northward over western Cuba, the notice told readers, and it was eastbound, heading for the south Florida coast.

In other words, it wasn't a hurricane. And it was heading away from Galveston.

The notice was simply signed "Moore." That was Willis Moore, director of the Weather Bureau and Isaac Cline's boss. When he filed that entirely erroneous notice regarding the track of a disturbance coming out of Cuba, Willis Moore was already deeply and strangely involved in what was about to happen to Galveston.

Accurate long-range tracking of hurricanes was hard to come by in 1900, of course. Still, Director Willis Moore's notice from the Weather Bureau in Washington, placed in the Galveston paper on the fifth, was so exactly wrong, about both the nature of the storm and the storm's direction, that it seems to suggest that both meteorology and international communications remained in a primitive state. Nobody, one might assume, ever knew anything in advance about a hurricane's strength or track.

But that's far from the truth. As early as Sunday, September 3, the storm was under the observation of meteorologists in Cuba. They were perhaps the best in the world—especially when it came to assessing and predicting the tracks of hurricanes.

The storm had grown by then into an unmistakably violent one. The Cuban forecasters knew it was a hurricane and which way it was really heading. And the United States and Cuba were intimately connected.

So how could the U.S. Weather Bureau have lacked good information about the real nature of this storm that it was blithely calling only a tropical disturbance? Or about that storm's likely track toward the Texas mainland? Why, as late as September 5, was the bureau advising people that a mere disturbance was heading away from the Texas mainland?

The grim answer to those questions has to do with a highly problematic relationship between the United States and Cuba. It reflects the overwhelming self-confidence of the men of the U.S. Weather Bureau. And it throws strange light on the career of the man who signed that notice in the *Galveston Daily News*: Isaac Cline's boss, the U.S. Weather Bureau chief, Willis Moore.

As it approached Cuba, the storm that had begun off the Cape Verde Islands and then grew in a fold created by two opposing winds had changed dramatically.

Days earlier, far eastward, two ships' captains had noted in their logs the unexciting presence of a tropical wave at sea. Yet probably by the time it entered the Caribbean Sea on August 30—crashing with thunder, jagged with lightning, and blowing with force—and certainly by the time it dumped heavy rain on the island of Antigua on September 2, this storm was no longer a single storm. It had grown into a group of storms, tangled up with one another. It was now what meteorologists call a tropical depression.

A tropical wave is unorganized. Its zone of low pressure travels on east-to-west winds across the ocean. It can be dissipated by other forces.

A tropical depression, by contrast, is a well-organized system of storms, circling around a central point. Much less susceptible to being broken up—since it's a revolving circle, resisting entry—a depression fosters low pressure in its center, and it grows rather than dissipates.

Caught up in one another, unable to let go, the multiple storms that made up this depression weren't like those afternoon storms that arise daily in the summer on the African grasslands. They draw moisture from the earth and then release it all back as rain on that same ground. This storm system, by contrast, was drawing energy from a variety of places all at once—hot air from the sea, the winds at its back, the winds it was creating—and it was dispersing energy—buckets of rain, claps of noise, jagged lightning—in many directions. On the move for many days, it wasn't about to stop now.

The system's rotation was counterclockwise. It was twirling in response to the rotation of the Earth. Thanks to centrifugal force, any object may veer to the left or right when traveling across a spinning surface, and in this case the object was a storm system; the spinning surface was that of the Earth, including the Earth's atmosphere. On the upper half of our globe, the Northern Hemisphere, objects responding to the Earth's rotation—when few other factors are in play—tend to turn counterclockwise. In the Southern Hemisphere, they tend to turn clockwise.

So it was that this tropical depression came circling into the sky above the Caribbean. It brought hard winds, torrential rain, and thunder and lightning. A depression like this is only the weakest of three kinds of tropical cyclone. The next strongest we call a tropical storm.

After that: hurricane.

On September 3, Antiguans came outside. The drenching storm system had passed. High winds had died. The air grew still but did not clear and dry out.

A heavy, still humidity prevailed. That's one effect of the passage of a tropical depression.

Then, in Jamaica, miles of railway roadbed were washed out.

The storm marched on. Next in its path: Cuba.

And as it came toward Cuba, the storm was confronted by Father Lorenzo Gangoite, a Jesuit priest. Father Gangoite, one of Cuba's great weathermen, knew exactly what he was seeing. Soon he would know exactly where the storm was heading. He would try to give urgent warning to the Gulf Coast of the United States.

But Willis Moore of the U.S. Weather Bureau disdained the Cuban priest and his ilk. In that emotion lay the seeds of a disaster for Galveston that now lay only days away.

As the head of the Belen Meteorological and Magnetic Observatory in Havana, Father Gangoite knew pretty much all there was to know about how tropical depressions develop into hurricanes. And when this depression hit the southwest coast of Cuba on September 3, 1900, he had a lot riding on his observations of it, and on what he could predict it might do. Gangoite hadn't yet shown he could nail such a prediction. He'd only taken over the observatory in 1893, and he had big shoes to fill.

In Cuba, the science of meteorology had been brought to a fine point by Gangoite's predecessor, Father Benito Viñes.

Meteorology, like much other science in Cuba, was the province of Jesuit priests. They'd developed an advanced body of knowledge and interpretation regarding bad weather. The Belen Observatory, founded by Father Viñes in Havana in 1858, was an extension of a Jesuit preparatory school, itself founded only four years earlier. At both the school and the observatory, Cuban priests under Father Viñes carried on the long Jesuitical tradition of inquiry, experimentation, publishing, and teaching.

There couldn't have been a better place from which to study and learn to forecast bad weather than the city of Havana. Its tropical vegetation, wrought-iron balconies, and painted stucco houses and porches were routinely subjected to violent downpours and torrents of destructive wind. One year, a hurricane removed the observatory's entire zinc roof.

Meteorology had thus became one of Cuba's most important sciences. The Jesuit observatory was perhaps the most advanced in the world. Father Viñes enjoyed celebrity for the amazing accuracy of his predictions.

Like Isaac Cline in Texas, Father Viñes in Cuba hoped not only to advance meteorological science but also to aid the progress of humankind. He soon made the small Havana observatory the hub of a forecasting network for the entire Caribbean Sea. He started a storm notebook full of descriptions of clouds, cross-referenced to instrument readings. He jotted down snippets of conversations with ship captains. He brought in telegraph reports and newspaper clippings.

From these data, Viñes created a system for understanding storm formation and making predictions. He published it all in newspapers so that ordinary people could understand and respond.

But his real genius lay in interpreting the meaning of clouds and how they related to hurricanes. Cirrostratus clouds are high and gauzy and are composed of ice crystals. They give a kind of cover through which a haloed moon may be seen or from which hazy sunshine emanates. Viñes realized hurricanes tend to produce these cirrostratus clouds—but only on the outer edges of a system. He therefore began to suspect that those clouds are created by winds flowing off a hurricane system miles high.

So if you were to see cirrostratus clouds in the tropics, Father Viñes deduced, you might really be seeing the farthest outer edge of a hurricane, which you wouldn't otherwise have any idea was out there. Because hurricanes are so massive—hundreds of miles across—the far outer edge that you're viewing may lie many days' travel away from the storm's deadly eye.

You know a hurricane is coming. And you still have time to act.

But not all forms of cirrostratus cloud signal the approach of a distant hurricane. The clouds must come in a certain specific shape: plumiform. That is, they appear to spread across the sky, fanning upward in plumes that seem to be reaching out from a central point. The bottoms of these elongations, Viñes further deduced, are pointing directly at the eye of the hurricane that produces them.

So now you also know the direction from which the hurricane is coming.

Using those theories, Father Viñes built a model by which meteorologists could ascertain that a hurricane had formed, calculate roughly how far away it was, gauge how fast it was moving, and even closely track its path. Then, like Isaac Cline in Texas, the priest put his method to the test.

In September 1875, Viñes issued a public hurricane warning based on telegraph reports from Puerto Rico passed on by the Spanish navy. Ships must not sail east or north from Havana, the priest advised.

Most captains followed instructions. The one who didn't was the captain of a U.S. steamer. He tried to beat the storm into the Straits of Florida. All hands on that ship went down when the eye of the storm passed northeast of Havana—exactly where and when Viñes had said it would.

The next year, Viñes repeated the trick—but this time he predicted a hurricane by using only his observations of cloud formation and his readings of barometric pressure. The storm blew in right on schedule.

Soon he had a telegraphic network of storm observers working the entire Caribbean, integrating reports from Cuban train operators, consular officials on colonial islands, and the U.S. Army Signal Corps, scattered throughout the region (the U.S. service was then still part of the Signal Corps). Viñes upgraded Belen, traveling to England to buy and calibrate new high-quality instruments at Kew Observatory. By the late 1880s, his telegraphic network integrated reports from every kind of colonial and independent government: Spanish, British, French, Danish, Dutch, Dominican, Venezuelan, and American. Everything about Caribbean weather went through Father Viñes in Havana.

When Father Gangoite took over the observatory in 1893, his new directorship coincided with a climax of political turbulence for Cuba—and especially for the island's relationship with its gigantic neighbor just ninety miles to the north. That political turbulence would have a strange impact on Cuban forecasting. And it would have deadly ramifications for Galveston in 1900.

The year before Father Gangoite took over Belen, José Martí, the Cuban writer and revolutionary, took his lifelong struggle for Cuban independence from Spain to a climax. Martí founded the Cuban Revolutionary Party, which brought together a wide range of radical groups that hadn't been able to work together before. As in Mexico eighty years earlier, the moment seemed ripe for a move in Cuba against the Spanish colonial rulers.

Some of the Cuban revolutionaries looked—if with mixed feelings—to the United States for help in gaining independence. And Americans, in turn, gazed—if sometimes greedily—on the feisty little island. Americans related Cubans' desire for independence to the heady days of the American founding. They also saw much economic potential in Cuba.

This suggested to some that the United States should simply annex the island. It was an old idea. As early as 1805, President Thomas Jefferson had considered taking Cuba. In 1823, President James Monroe included Cuba in the Monroe Doctrine, where U.S. hegemony in the Western Hemisphere was first proclaimed. Some early private "filibusters" had tried and failed to capture Cuba. In 1891, the *Detroit Free Press* put it this way:

> *Cuba would make one of the finest states in the Union, and if American wealth, enterprise and genius once invaded the superb island, it would become a veritable hive of industry. . . . We should act at once and make this possible.*

While the Cuban revolutionary Martí feared annexation, he admired the liberties set out in the U.S. Constitution. He believed that by helping Latin American independence movements, the United States might redeem what he condemned as its inherent racism and political corruption.

In Key West and Tampa, Martí raised both money and revolutionary troops. He called for a multiracial uprising in Cuba. He wanted to place blacks, whites, and natives on equal terms, ending slavery and adopting civil liberties.

Two years after Father Gangoite became the chief meteo-rologist at Belen, the insurrection broke out in Cuba. Martí, the great voice of Cuban liberty, was killed in that war. Yet so successful was the uprising he inspired that, by 1897, Spain began trying to come to terms with the Cubans by liberaliz-ing the colonial government. By then the revolutionaries were in control of much of the island. There was an autonomous Cuban government in charge of Havana, where Father Gan-goite went on watching the skies and seas.

No deal, the revolutionaries said. Spain must go.

Public opinion in America, meanwhile, was excited about rev-olution in Cuba. Old ideas about straight-up annexation were mingling with new notions about expanding American influ-ence, if not literal sovereignty, into the Caribbean. Interven-ing in the Cuban revolution, and opposing Spain in support of democratic change, became a war cry among some in Con-gress and the press.

At first, President McKinley refused to intervene. He de-manded only that the Spanish colonial government stop driv-ing Cubans from their homes and start negotiating with the rebels. That frustrated the U.S. interventionists.

But they got a new chance when Spanish loyalists rioted against the new government in Havana. That's when the United States sent its battleship USS *Maine* into Havana harbor. Os-tensibly, the ship was there to protect the lives of Americans living in that city.

On February 15, 1898, the *Maine* exploded and sank in the harbor. Nobody knows why. But the explosion was enough. William Randolph Hearst, owner of the *New York Journal*, was bent on getting the United States into a war with Spain. He was also carrying on his own war, against the other big newspaper publisher, his archenemy Joseph Pulitzer of the *New York World*. Headlines in the two papers screeched, at competing volume, that Spain was responsible for the sinking of the *Maine*.

Dramatic speeches in Congress persuaded business leaders, formerly skittish, to support intervention. McKinley's own assistant secretary of the navy, Theodore Roosevelt, raised the nattily dressed volunteers known as the Rough Riders. The president's stance against intervening in the Cuban War for Independence began isolating him from Congress, from the press, from the public, even from many of his own appointees.

In April, at the president's behest, Congress declared war. The United States blockaded Havana.

Meanwhile, the Jesuits at the Belen Observatory continued their punctilious scientific work on hurricane prediction. It's not that Jesuits were nonpolitical. Anything but: Father Viñes, Gangoite's mentor, had come to Cuba after fleeing Spain, with much of the Jesuit order, after that nation's Liberal Revolution of 1868.

But Cuban Jesuit scientists tried to ignore colonial politics. When Cuba's Spanish governors repeatedly closed down secular scientific institutions amid the long period of unrest, meteorology stayed with the Jesuits.

Cuban weathermen couldn't stay untouched by politics forever, however. Forecasting requires constant observation of phenomena that lie well beyond human politics. But in 1900, U.S. support for the Cuban rebellion, and the joint success enjoyed by the American government and the Cuban patriots, had political effects that played into a terrible natural disaster.

When Father Gangoite was considering the likely path of the storm that was drenching Cuba on September 3, 1900, he wanted to prove his own expertise. But he also wanted to show that his predecessor, Father Viñes, hadn't been working some magical personal mojo in sniffing out hurricanes. Gangoite wanted to demonstrate that Viñes's models were scientific. The results could be repeated.

So Gangoite observed this new storm. He saw that it was changing fast. It was twirling on its own axis as it zoomed across the spinning Earth—yet it hadn't formed that perfect, and perfectly deadly, spiral form that we associate with a hurricane. There wasn't yet an eye of low pressure at the system's center. Its winds, while hard and rough, still did not reach above 60 miles per hour.

The storm nevertheless already had the power to knock down buildings and wash away train tracks on Jamaica, Cuba, and other islands. The violent cloud mass was turning and turning, and its winds were blowing in a circular form, even as the whole system was moving on its northerly and westerly course over and then away from Cuba.

If a storm like this were to follow the path that Father Viñes's rules had always correctly predicted it should, it would soon be a hurricane. And Father Gangoite thought he could tell exactly where that hurricane was going to go: toward the coast of east Texas.

If Gangoite's predictions had been heard in the United States, there would have been time to give the people of Galveston and the Texas mainland a fairly long-range warning of the impending destruction. They might not have believed it, of course. Ordinary citizens of Galveston—from Daisy Thorne to Chief Ketchum, from Boyer Gonzales's friend Nell Hertford to Annie McCullough to Arnold Wolfram—would have had no direct evidence that a deadly storm system, having left Cuba on Wednesday, September 5, 1900, was moving out into the Florida Straits and heading for them.

Captain Halsey of the *Louisiana* would soon meet the storm in the straits. But nothing was going on in Galveston to suggest that trouble lay ahead. Many citizens there laughed down storms anyway.

It didn't matter. They never heard Gangoite's forecast anyway. Both parts of that two-part forecast—that the storm was about to become a hurricane, and that it would turn westward, toward Texas—were anathema to the U.S. Weather Bureau. As far as the bureau was concerned, those two things could not happen. No Cuban was going to tell Americans that they could.

Resentment and disdain for Cuban forecasting had become an entrenched conviction at the U.S. Weather Bureau by the fall of 1900. The man in charge of tracking and reporting from Washington, D.C., was Willis Moore, head of the bureau—he had succeeded General Greely, who had made Isaac Cline head of the Texas section—and he made squelching Cuban forecasting one of the most important reforms he brought to the office.

When he took over the bureau in 1895, Willis Moore was thirty-nine. Filled with careerist energy, and under pressure from his boss, Secretary of Agriculture Julius Morton, he was on a tear to make the bureau a new model of efficiency. In the recently completed Weather Bureau Building, on the corner of M and Twenty-Fourth Streets, not far from Rock Creek in Washington, Moore presided over a thoroughgoing revamping of the service.

First, he started tracking the veracity of each U.S. forecaster. He set up a contest: a group of selected forecasters had to make predictions for the same city—not their own—based on the same set of daily and hourly reports. This kind of thing put the observers under such intense pressure, Moore bragged when testifying before Congress, that Weather Bureau workers were committed to insane asylums more often than workers in any other agency. With Moore in charge, Congress was getting its money's worth.

Perhaps most important, Moore tightened the rules concerning local forecasting—especially regarding storm warnings. Observers in the various state and federal weather stations had often taken it on themselves to issue local warnings when storms seemed imminent. Sometimes the area's weatherman would pass the warning on to the local paper; often, especially near a coastline, he would hang flags, a memory of the old Signal Corps days.

Moore believed local weathermen had been over-warning the public. There was a tendency to sow panic. It created an unhappy impression that the bureau was not fully in control.

Now all storm warnings, from everywhere in the country, had to go to Moore at his hub in Washington. The local weatherman cabled his hourly, daily, and other regular reports to the central office: each weather station in the United States, Mexico, and the Caribbean sent temperature, atmosphere, and wind conditions by telegraph. Weather clerks in Moore's office aggregated the morning data into a national weather map; the map was then telegraphed back, in turn, to each station. All warnings were issued in that manner. It was for Washington, not for local weathermen, to determine what was going on locally.

And for fear of panicking local populations, Moore banned certain words from all official weather reports. He banned the word "tornado." And "cyclone." And "hurricane."

Moore's new way of running the U.S. Weather Bureau clashed, not surprisingly, with the Cuban forecasters' methods for tracking and predicting hurricanes in the Caribbean Sea. The Cubans used the word "hurricane" freely—and their predictions had long traveled through telegraph weather networks in which the United States also participated.

In Cuba, first Father Viñes and then Father Gangoite sat at the hub of reporting that overlapped with mainland U.S. systems on the Gulf Coast. Moore wanted to change all that.

In 1900, Willis Moore got his chance. When the United States blockaded Cuba and then invaded the island in support of revolution against Spanish control, Moore and his boss, Secretary Morton, were immediately at President McKinley's side. The U.S. military operation in Cuba needed weathermen.

Moore showed the president his weather maps. He offered expert advice for getting the U.S. Navy around Caribbean storms. McKinley, impressed, ordered Moore to set up a series of storm-listening posts in Mexico, in Barbados, and in the Caribbean. To establish those posts, Moore picked what he considered his best men. He put Isaac Cline in charge of the Mexican networks.

But for the West Indies, unfortunately, Moore assigned Colonel Henry Harrison Chase Dunwoody. An old Signal Corps officer, Colonel Dunwoody had made his name by scoffing at the value of meteorological science in making predictions. He'd played a role in discrediting and undermining probably the top American meteorologist of the day, Cleveland Abbe. Now in charge of weather forecasting for the Caribbean Sea, Dunwoody was ideally positioned to continue his crusade against the uses of science in tracking storms—even while advancing his own career in weather.

The Cubans were the worst offenders, Moore and Dunwoody agreed, because they pretended that hurricanes could be predicted. And the two U.S. weathermen now had the power to put that pretense to an end.

By the fall of 1900, the United States was in control of Cuba. The United States hadn't annexed the island nation. That was prohibited by an amendment to Congress's declaration of war. Still, the United States did exercise effective control. With the victory of the Cuban revolutionaries, Spain had agreed to leave the island for good. Yet even as Cubans got ready to develop a constitution and hold elections, the United States set no timetable for withdrawal. No Cubans attended the peace talks with Spain held in Paris. While the resulting treaty, signed in December 1898, made Cuba officially independent, General William Shafter, the U.S. conqueror, barred Cuban rebel forces from the surrender ceremonies held in Santiago de Cuba. A U.S. governor, General John R. Brooke, ran the island.

In September 1900, when the island was hit by the tropical depression that Father Gangoite was observing, the first Cuban elections had just taken place, back in June. But in those elections, only local mayors, treasurers, and prosecutors were chosen. Elections for delegates to a representative general assembly were scheduled for September 15, and, thanks to rules set by Governor Brooke, the electorate was tightly restricted.

So there was not, in September 1900, a politically independent Cuba. The U.S. government administered the island. That gave Willis Moore in Washington and Colonel Dunwoody in the Caribbean a means of silencing Cuban hurricane forecasting.

Moore and Dunwoody's problem with Cuban traditions in weather forecasting was that the forecasts seemed hysterical and primitive overreactions to weather—despite those traditions' extraordinary history of accuracy, based on Jesuitical empirical and experimental science. To the Americans, Cuban knowledge was nothing but the superstitious lore of a backward people, lacking the Yankee grit and know-how that was making America a great leader on the world stage.

And nobody could forecast hurricanes anyway. As Colonel Dunwoody put it, the sources, progress, and ultimate courses of hurricanes might as well be "a matter of divination."

So Moore and Dunwoody appointed one of their own to assert a big, strong, guiding American presence in Cuban forecasting. That appointee was William B. Stockman, a veteran of the bureau going back to the Signal Corps days. Under Dunwoody and Moore, Stockman set up shop in Havana and took charge of all of the U.S. weather stations in the region.

In one of his early reports to Moore, Stockman simply eradicated the entire history of the Cuban weather networks. He told Moore that Cubans had never heard of such a thing as forecasting. The locals were "very very conservative," Stockman reported, ". . . and forecasting the approach of storms, etc., . . . was a most radical change." Fortunately, the United States was here to set things straight.

Stockman may have actually believed what he was saying. He may have been entirely ignorant of the advances made at Belen; he may not have known about Father Viñes's networks.

Or maybe Stockman was a smart careerist. He knew Moore and Dunwoody would take happily to the notion that the Cuban weather service, which had so often been right, didn't have any real expertise.

In any event, with the United States in charge of Cuba, and with Moore's Weather Bureau dazzled by its own prowess in rationalizing the locals' primitive systems, it was especially important, Stockman advised Moore, that the bureau not be guilty of causing "unnecessary alarm among the natives."

And there was yet another problem with the Cuban weathermen. Father Gangoite's Belen Observatory in Havana, Stockman claimed, had been secretly piggybacking on U.S. reports. Agents in New Orleans, he reported to Moore, stationed at the College of the Immaculate Conception there, nabbed copies of the daily weather maps coming out of Washington. The agents then sent the U.S. maps by undersea telegraph line to Havana. Such shifty shenanigans allowed the Cubans, as Dunwoody put it, "to compete with this service."

In other words, according to Stockman and his bosses Dunwoody and Moore, the Cubans never got things right, but when Cubans did get things right, it was because they stole U.S. data. Having pinched good reports, the Cuban forecasters whipped a silly, uneducated, overemotional population into a frenzy with overblown warnings of monster storms.

As far as Director Moore was concerned, all this had to stop. In late August of 1900, just as a certain tropical wave was forming off West Africa, he decided to put his foot down. He would deal once and for all with these Cuban annoyances.

Hurricane season was well under way. This was the perfect time, Moore calculated, to shut down all communication between Cuban weathermen and the people of the United States.

It would take some string pulling. Fortunately for Moore, the U.S. War Department controlled all of Cuba's government-owned telegraph lines. Those were the same lines over which Father Viñes had established his fabled hurricane-warning system, not for Cuba alone but for the entire region. That system was about to become a thing of the past, if Moore had anything to say about it.

And he did. From Washington, Moore contacted Dunwoody, his top operator in the West Indies. Moore ordered Dunwoody to ask the U.S. War Department to formally ban from Cuban government telegraph lines any and all messages referring to weather.

The War Department responded quickly. Martial law effectively prevailed in Cuba, so the weather-telegraph ban went into effect in the last week of August. That same week, ships' captains out in the Atlantic, not far north of the equator, first sighted the storm that would drench Cuba and head toward Galveston.

But Moore went further. Even among U.S. stations there must be no direct communication. No weather information in the U.S. Weather Bureau's office in Havana could travel to the office in New Orleans. That was in keeping with the director's long-standing preferences. Despite the proximity of Havana and New Orleans, everything from Havana must go directly to Washington. Washington would filter the Havana reports and decide what information to give New Orleans and the rest of the Gulf Coast.

Moore went further still. He reached out to Western Union, the commercial telegraph company. That company's lines weren't under government control. Moore wanted to ban private Cuban weather-related cables as well.

He knew he couldn't demand that Western Union literally censor private messages. But he could ask the company to manage what a later age would call bandwidth. He requested first priority for U.S. Weather Bureau transmissions. Next would come any non–weather-related messages. Cuban weather messages should get the lowest priority.

Western Union showed a patriotic willingness to cooperate with the government. It gave all non-weather content superior privileges. Any private telegraphs from Cuba to the United States regarding weather would be slowed, bumped, and in many cases, Moore hoped, discarded. His blackout of Cuba was almost total.

Father Gangoite had correctly predicted the arrival of heavy rain in Cuba. He had called those rains, correctly, "a cyclonic disturbance in its incipiency": that is, a hurricane in the making. ("Cyclonic" was just the sort of word the U.S. bureau frowned on.) And he'd predicted—again correctly—an increase in violent force as the storm left the island and entered the Straits of Florida.

And yet thanks to the triumphant dictate of Willis Moore and the U.S. War Department, accomplished only days earlier, banning direct weather communication from Cuba to the United States, Gangoite's predictions could not legally travel off the island. With the storm now raging in the straits and moving quickly toward the United States, only Washington could tell Galveston anything.

That week, Isaac Cline, along with Joseph Cline and another assistant, John Blagden, went about their daily rounds in Galveston. On the roof of the Levy Building they took readings.

They looked out across the vast, still, blue gulf sky. They observed the characteristically mild surf on the gulf beaches.

They made their periodic and daily reports to Washington. They made and considered daily weather maps based on information from Washington. Nothing looked out of the ordinary.

The bureau in Washington did, of course, report to Galveston on the progress of the disturbance coming out of Cuba. "Not a hurricane," Moore called it (evidently you could use the word as long as you put "not" in front of it). The course of this non-hurricane, as the bureau saw the situation and reported it back to the gulf stations, would not affect Galveston. The storm would instead go into a classic "recurve." This recurve effect was among the many laws of storms, according to the United States. Exiting the Caribbean on a northerly trajectory, storms simply cannot, these laws held, continue on a northwestern track. A storm thundering out of Cuba over the Florida Straits must turn toward Florida.

The inevitable turn toward Florida was a good thing, the Weather Bureau believed. For a number of other predictable things would then have to happen. Arriving at the Florida peninsula, the storm would start sweeping a surface turned not horizontally against it, like a wall, as on the Gulf Coast and its offshore islands. The storm would instead be coming up against the vertical orientation of the peninsula. Broken coastline on the Florida gulf side would prevent the storm from hitting any land mass with head-on force.

So as this supposedly mild storm coming out of Cuba continued on its supposedly northerly path, it would have to turn east, and then it would lose what little power it had.

That was the official U.S. prediction. The system, the bureau telegraphed New Orleans, was "attended only by heavy rains and winds of moderate force." There would be rain and high winds along the Florida coast, with some damage to moored ships and shoreline property. The storm would then move northeastward, through the southeastern states, weakening as it went. New Orleans and points eastward were authorized to hang the red-and-black storm-warning flags, letting captains know of moderately disturbed seas. (Captain Halsey saw one of those flags as he crossed the bar at Port Eads, Louisiana.) But any residual action in the gulf would quickly dissipate. And no warnings were in order west of New Orleans.

The storm would probably, the bureau informed the gulf stations, "be felt as far northward as Norfolk by Thursday night and is likely to extend over the middle Atlantic and South New England states by Friday." It was moving away, that is, and heading for the East Coast. After that, the storm was expected to exit the United States into the Atlantic somewhere in or above New England.

There was another well-regarded meteorologist in Havana: Julio Jover. Like Father Viñes and Father Gangoite, he had a reputation for accurate storm prediction during hurricane season. Jover was, if anything, even more of an irritant than his Jesuit colleagues to Stockman, Dunwoody, and Moore.

Now, regarding the storm out in the Florida Straits, which had just drenched and battered his island, Jover was making his own prediction. It dovetailed perfectly with Father Gangoite's ideas. Together, the two major Cuban weathermen were offering a forecast that was nearly a perfect opposite of the forecast made by Washington.

As early as Wednesday morning, when the storm was still leaving Cuba, Julio Jover said this: "We are today near the center of the low pressure area of the hurricane."

He'd said it: "hurricane." And he mailed that forecast to the Havana newspaper *La Lucha*. He was infuriated by the cable ban. The mail was all he could use. But Jover was at least on the record.

Father Gangoite meanwhile referred to Father Viñes's rules of storm travel and made some notes of his own on what he was observing. Late Wednesday night, there was a big halo around the moon. The halo did not dissipate. At dawn, the sky turned red—deep red—and "cirrus clouds," Gangoite said later, "were moving from the west by north and northwest by north, with a focus on those same points." To him that meant the storm had transformed drastically after beating up Cuba.

The change in the storm took three forms. First, the storm had gained intensity, as Gangoite had predicted it would.

Second, it had gained structure. No longer what we would call a depression, the whole gigantic thing had begun twirling around a more definite eye. That was a hurricane.

These Cuban weathermen couldn't directly see the hurricane. They only had observation and deduction, based on years of study. And what they saw in imagination is exactly what Captain Halsey met, in fact, at sea: a huge hurricane in the Florida Straits.

Gangoite's third note was the most important. The hurricane was taking a path very different from what Willis Moore predicted—the reverse, in fact. Both the prevailing winds in Cuba and Father Viñes's rules for observing hurricanes suggested that, far from recurving northeastward toward Florida, as the Weather Bureau had it, the hurricane was actually heading northwestward, straight for the Texas coast.

The Cubans even pinpointed the destination of the storm's center. It would go somewhere between Abilene and Palestine on the Texas mainland. Lying right in that path: Galveston.

They would soon see, as Father Gangoite put it with growing frustration, "who is right." With the telegraph ban silencing the Cuban forecasters, there was no way to warn U.S. weather stations in New Orleans or Galveston of what these men knew was about to happen.

Jover and Gangoite could do nothing but wait, in outrage, for disaster to strike Texas.

Willis Moore had blocked the forecast. But he couldn't stop the hurricane.

PART II

MAELSTROM

CHAPTER 6

GALVESTON: THURSDAY, SEPTEMBER 6

WHAT MADE THE STORM SO QUICKLY APPROACHING GALVESton a hurricane—whatever Willis Moore said—and no longer a tropical depression or tropical storm? And why did it not, in this case, "recurve"—Moore was actually correct in believing that many hurricanes do—and head toward Florida?

The answer to both of those questions has to do with features of air pressure. The force of air, that is, measured by a barometer, which makes us feel good when it presses on us fairly hard, and makes us feel sluggish when the pressure lessens.

As this particular swirling storm system moved northward across the Florida Straits, it was still spinning counterclockwise, thanks to the rotation of the Earth. And by now it had become truly gigantic: a fully organized, circular system of destruction, literally hundreds of miles across. And that whole swirling mass of violent rain, pushed hard by relentless winds, was letting off energy in wild spirals of drenching thunderstorms.

At the system's center there was now an eye. The spinning complex of storms was turning around a large, roughly circular space, empty of any evident action. The eye of this hurricane moving toward Galveston might have been thirty or more miles in diameter. In the middle of the maniacal turbulence, the axis at the center of the eye resembled a state of pure stillness.

That doesn't make the eye of a hurricane any place you'd want to hang around, supposing you could somehow rise into the eye and spend some time suspended there outside an airplane. The stillness at the axis is caused by drastically low air pressure in the eye. Around it rises a "stadium effect": the high wall of thick cloud that defines the eye. If you could get up there and inside it, you'd think you'd entered a world ringed by snowy mountains that rise far across a snowy plain, dozens of miles wide in every direction you look.

The thickness of those far-off mountainous clouds surrounding the eye make the word "cloud" seem wrong: it's an "eyewall," in meteorological terms. Despite the eerie stillness within the eye itself, the wall that rings it serves as the location for the most violently swirling thunderstorms the hurricane can bring.

And the air inside the eye is horribly hot and strange in human terms. It doesn't rain, but the air pressure is so low that the stillness wouldn't seem peaceful but painfully oppressive. The air is so dry that it would feel rough, inhospitable. There's nothing familiar or beneficial to human beings about any component of a hurricane, especially the eye.

One of the effects of the otherworldly low pressure in the vast eye of a hurricane is that the storm doesn't lose power as it travels but actually gathers up immense energy. As this storm traveled toward Galveston, it drew heat, evaporating from the hot gulf waters, into its central zone of low pressure. The heat fueled the storm, turning it faster, causing it to throw off new and harder winds, even as it was pushed by winds at its back.

The monster could only grow. It would soon become what today we call a category four hurricane. The highest number on that scale is five.

There also happened to be, at that moment in September of 1900, a big zone of high pressure sitting well to the east of the storm. This high-pressure zone bordered the Florida Keys— that string of narrow islands curving southwestward from the southern tip of the state's long peninsula.

This high-pressure zone to the east of the Keys caused an exception to the so-called rule of hurricane recurve, which Willis Moore thought was immutable. A recurve would have drawn the hurricane eastward to Florida and then up the East Coast. Many storms traveling north from Cuba into the Florida Straits do indeed tend eastward, as Moore's men were sure this one was doing.

But in this case, the storm was actively blocked in that direction. The high pressure at the Keys pushed it away. High pressure, with its tightly packed molecules, can hold off low pressure; low pressure, like that at the eye of the hurricane, seeks still lower pressure. Winds off the Keys, blowing from east to west, added to the pushback.

So even as it grew into a monster of disastrous violence, the cycling hurricane's eastern edges were pushed away from Florida, brushing against and bouncing off the high-pressure zone there. That action kept the whole system—in defiance of all U.S. forecasters' rules—from recurving eastward. The hurricane was cycling steadily toward the Texas coastline, just as the Cuban forecasters had predicted it would.

On it went. Its world was titanically violent. Drawing new energy constantly from the hot sea below, pulling those waves high upward, throwing wind in every direction as it circled, unleashing gigantic thunderclaps and streaks of jagged lightning, and pouring hard rain, the storm—blocked from any possible turn eastward—was invited west-northwestward by low pressure there, pushed that way by winds at its back.

The track had now become unchangeable. That track put the hurricane farther and farther away, every second, from the track that the U.S. Weather Bureau was drawing, with such infinite confidence, care, and precision, on the national weather map.

At 6:00 on Thursday morning, September 7, the people of Galveston, Texas, were rising, looking forward to the weekend, and hoping for relief from the heat. That's when William Stockman, Director Moore and Colonel Dunwoody's man in Havana, filed an official observation from Cuba. Regarding the storm that just had beat up Cuba, Stockman described its current position as 150 miles north of Key West—the southernmost point of the Florida Keys and of the United States as a whole.

That in itself wasn't exactly wrong. As Captain Halsey and his crew and passengers would shortly find out, the storm had indeed passed Key West the day before, clanging against the high-pressure zone to its east. That action caused intense winds on the Keys, and those winds quickly become gale force from Key West to Tampa.

The barometer fell to the lowest level ever recorded in Key West. The weathermen's telegraph wire, linking Key West to Washington, D.C., blew down.

It was also true, as Stockman reported Thursday morning, that the storm was continuing northward. But it was continuing northward with a critical difference. Stockman reported not only that the storm was north of Key West but also, based on the recurve rule, that it was north by east.

That was exactly wrong. The storm was north by west of the Keys. It was traveling across the Gulf of Mexico toward the coastal mainland to the west. It was starting to raise the gulf seas to heights that had never before been recorded. Just as the hurricane was building on itself, changing water and wind in ways that fed back into the storm and amplified its own destructive energy, Stockman's erroneous tracking fed into a growing set of assumptions, and then into some new miscalculations, made by the U.S. weathermen.

Now came the U.S. Weather Bureau's forecast from Washington. At 8:00 that Thursday morning, the bureau confirmed its prediction, telegraphed to New Orleans and Galveston the day before, regarding the storm recently reported in the West Indies. This storm, the bureau reported, was diminishing in power and had recurved, as such storms always must. So it had arrived in Florida—just as expected.

At Key West, the bureau went on to report, the wind had quickly dropped from gale force to the lightest kind of breeze. Then it had picked up and changed direction. The wind had been blowing from the northeast; now it came out of the south.

That meant to Director Willis Moore that the storm must be proceeding on the track they'd already determined it must be traveling on. They didn't read the wind change as indicating a zone of high pressure in the Keys, which would push the storm off to the west.

So this "storm"—certainly not a hurricane!—would "probably continue slowly northward and its effects will be felt as far as the lower portion of the middle Atlantic coast by Friday night," the bureau told its stations on Thursday morning. Not a hurricane, and heading northeast: that's the precise reverse of a hurricane warning for the Gulf Coast.

In fact, some fisherman on the New Jersey shore, having received the national report, cabled Director Moore for advice in dealing with this storm that was supposedly heading their way. Never one to hesitate, Moore cabled right back. "Not safe to leave nets in after tonight," he warned them. A rough storm—though by no means a hurricane—was heading right for the Jersey shore, Moore was certain.

If Boyer Gonzales, scion of the Galveston Gonzaleses, now painting in New England for the summer, had been reading the weather forecast in Thursday's afternoon Boston paper, he would have expected some heavy rains a few days hence. That might even have offered him dramatic conditions for capturing a certain northern marine light in his painting.

Boyer certainly wouldn't have had any reason to worry about his younger brother and his sisters, or about his friend and correspondent Miss Nell Hertford, all sweating it out back home in Texas.

In Galveston, meanwhile, everything looked fine that Thursday morning—if continuing still and humid—when Joseph Cline, Isaac's brother and chief assistant, went to the top of the Levy Building to take the morning readings.

Barometric pressure within the normal range. Light winds.

Temperature already 80 degrees early in the morning—hot, but slightly cooler than it had been, actually. The huge sky over the Levy Building and out to the calm gulf was as clear and blue as could be.

As usual, Joseph coded those readings and sent them via messenger over to the Western Union office a few blocks away. From there, the readings would go to Washington.

Thus the local Galveston readings too played into disastrously faulty tracking of the storm by Moore and his men in Washington. The clarity of the sky, as viewed from Galveston, the slight drop in temperature there, and the normal barometric reading: all of these would have suggested to U.S. meteorologists of 1900 an utter absence from the Gulf of Mexico of the storm that had drenched Cuba only days earlier.

In later years—partly as a result of what was actually happening that September in the gulf—those signs might just as easily be taken as predicting trouble: "the calm before the storm," as the saying goes. Even a normal barometric pressure would not suggest to meteorologists today that no storm could possibly be on its way. Pressure can fall far more quickly, and far more dramatically, than the observers of 1900 knew.

That's exactly what was about to happen in Galveston.

The other thing the Galveston weather crew did every morning, along with sending local observations to Washington, was relay the information that came from Washington to the Galveston Cotton Exchange. Along with Isaac Cline, his brother Joseph, and the second assistant, John Blagden, the weather station in the Levy Building had a man dedicated to that exact purpose, a skilled printer.

Every day, the printer made a graphic image of the national weather situation based on the data from Washington, using a system of codes to show high and low pressure and other conditions. Then, over at the hefty and ornate Cotton Exchange building, a professional mapmaker used a system of colored chalk to draw a big version of this daily map, so that all the men wheeling and dealing on the trading floor could see it as they loudly bid and sold at top speed. Growing conditions and transportation are subject to weather, so the prices of cotton and cotton futures were subject to weather too. On the big Cotton Exchange map, temperatures, wind speeds, and air-pressure readings were noted for the entire country.

A system of circles denoted clear or cloudy skies. Little arrows showed wind direction. Letters signaled other conditions: "R" for rain, for example.

On that Thursday morning, September 7, as the excitable men making deals on cotton looked up from the trading floor to squint at the weather map, Captain Halsey and his passengers and crew on the *Louisiana* were beginning their encounter with a giant system of calamitous energy. That monster was moving fast toward Galveston, only about a day away. But nobody looking up from the floor at the Cotton Exchange map could have imagined the screeching winds that rolled and pitched the steamship as it tried to ride out the worst hurricane its captain had ever experienced.

On Thursday afternoon, Isaac Cline himself took readings. He noted only scattered clouds and a fresh wind.

And at 1:59 P.M., Cline received a telegraphed report from Washington. The storm that had drenched Cuba was now, as expected, centered over southern Florida, the bureau reported with confidence.

But observers in central Florida, surprised to read that

report, now began a slow process of correcting the Weather Bureau's forecast. Meanwhile, in Galveston, Cline took readings that night. It was hotter now—just over 90 degrees. That might cause disappointment for the citizens awaiting the weekend. The wind was out of the north. The barometer was down—but just barely. There were scattered clouds.

Cline reported all of that in code to Washington and went home to bed.

CHAPTER 7

FRIDAY: THE WAVES

EVERYTHING HAD STOPPED MAKING SENSE.

It was Friday, fewer than twenty-four hours after Isaac Cline's untroubled Thursday-night readings at the Levy Building, and both Cline brothers found themselves deep in a state of anxious foreboding. They were scrambling to figure out what was going on.

First of all, the Weather Bureau had abruptly reversed itself. Friday morning, Isaac Cline received a telegraph from Director Moore. Moore ordered Cline to raise the red-and-black storm-warning flag. That was to alert ships' captains to expect trouble in the gulf.

Why, all of a sudden, this change? It wasn't only a change, but a total contradiction of everything the bureau had reported only yesterday. If that storm from Cuba was really now losing power across Florida, as reported the day before, heading up the East Coast and threatening to drench New Jersey and

New England, then this abrupt storm warning for the Gulf of Mexico was nothing but baffling.

At 10:35 that morning, Isaac Cline ordered the storm-warning flag hoisted. And he started thinking hard.

Cline didn't know it, but this is what had happened in Washington. The weathermen in bureau headquarters had begun getting surprising reports from local stations on the East Coast. The stormy weather predicted there had entirely failed to arrive. The winds that battered Key West did not start blowing in central Florida after all. Savannah and Charleston were not being drenched. Those fishermen in Long Branch, New Jersey, worrying about their nets had nothing to fear.

There was only one conclusion. The men in Washington finally drew it. The storm that had left Cuba on Wednesday must, after all, still be somewhere in the Gulf of Mexico.

But where?

Also: it must not be heading northeast. It must heading northwest.

Who is right? Father Gangoite had asked. His question was rhetorical, since Gangoite knew that he and his fellow Cuban forecaster Jover were right, and that Stockman, Dunwoody, and Moore were wrong.

On Friday morning, it began to become obvious to the U.S. Weather Bureau too that, regarding the direction and the placement of that storm, the Cubans were right.

And yet on one point, Director Willis Moore remained insistent.

The Cubans were correct about the storm's direction. He could accept that.

But the Cubans couldn't be right about this being a hurricane.

If the storm was really traveling from Cuba to Texas, it couldn't be a hurricane, Moore knew. Because hurricanes can't do that.

So this thing heading across the gulf, "slowly northwest," as one of Moore's cables finally admitted on Friday to his Gulf Coast weather stations, was only a modest storm system, he was sure. There would be high winds along the Texas coast on Friday night and on Saturday. Hard rain was probable too.

Then in Galveston things got stranger still.

That Friday afternoon, a heavy swell formed just off Galveston's long gulf beach. The swell was coming from the southeast. And it arrived with an ominous roar.

Yet there was nothing about the sky to go with this. None of the brick-dust reddishness of mist that had often presaged hurricanes before. There was nothing unusual about the clouds.

No red sky at morning? Then no real reason for warning. The barometer was indeed falling—but only very slowly. A storm was clearly coming from the southeast, as storms had come before. So the correction to the Weather Bureau forecast made sense after all. The clouds, meanwhile, were coming from the northeast.

So a severe storm, probably. The ominous roar of the swells confirmed it. There was nothing in the air, or on the instruments, to suggest anything more seriously amiss.

In fact, to the ordinary citizens of Galveston, untutored in the signs of storms, the warning flag flying from the Levy Building might have seemed preposterous. The heat seemed to be about to break. The heavens were blue, with pretty clouds lying along the horizon of the gigantic light-filled firmament that is the gulf sky.

If those clouds meant rain, good. A cooling trend would no doubt follow.

And those swells off the beach, multiplying now and rolling in, caused actual surf. To those living near the flat gulf, surf was something unusual. Surf was something fun. Friday became a nice day for a trip down to the gaudy pavilions at the bathing pier. Maybe take a trolley ride on the track that ran along the beach over the water.

Crowds began to gather on the beach. Bathers started having a good time with waves that came from somewhere a long way out and rolled high up on the beach. Soon those waves were jumping high enough to nearly touch the pier's electric lamps that stood tall above the water.

Few in Galveston had seen anything like that before. There had been storms, but this looked to be turning into an exciting one. On foot, on bicycles, and by carriage and wagon, people kept coming down to the beach.

The surf was fun, but a swell is a special kind of wave, and it portended no good for Galveston. Coming from somewhere far away, swells are caused by winds you can't feel directly, because they are blowing far from where you are. And they're blowing hard.

Local winds do, of course, cause waves. At ocean beaches, which normally have some surf in any weather, a nearby gust can lift surf into crashing breakers. On a gulf, bay, or sound beach, usually quieter, unusually hard local winds can create ocean-like surf. Even a wind blowing across a puddle of rain will lift the puddle's surface in ripples; winds blowing across lakes make big waves.

But a swell is a wave that has been lifted by a hard wind far out to sea. This wave has then traveled a long distance, arriving on a beach that may be having relatively calm winds. For swells to appear one after another on a beach, a far-distant wind must have been blowing very hard for a long time across a large expanse of water, creating long waves.

Unimpeded—as in the Gulf of Mexico—such waves can travel astonishingly long distances. Under the right conditions, a wave generated on one side of the world can organize itself during its long trip, minimize the kind of chop that sometimes breaks up big waves, and arrive days later, intact, on a beach on the other side of the world.

And as swells continue to be generated, many miles from the beach where they arrive, one will follow another toward that beach. The greater the distance they must cover, the greater their separation. Swells appear on the beach in succession, with pauses—sometimes very long ones—between them.

If you're standing on the beach, and if you're able to tune out a multitude of distractions—chop, normal wave patterns, excited beachgoers—and note the time that elapses between the swells' arrival, you can estimate your distance from whatever storm is generating them. And as those timings change, you can estimate the rate at which that storm is moving toward you.·

Isaac Cline hadn't yet started timing the swells bringing high crashing surf to Galveston's gulf beach. Cline and his assistants were busy all day. Even in advance of what they still believed would be nothing more than a severe storm, there was much for the weathermen to do. Detailed information had to be given out and constantly updated. Bathers and holiday-makers might delight in the thrill of the rising surf, but ship captains and businesspeople and cotton investors needed real information. It was up to the weathermen to get it to them.

So because they'd raised the storm-warning flag, and thanks as well to the crashing surf on the beach, Isaac and Joseph Cline had to communicate not only with the newspaper and the Cotton Exchange but also, by telephone and in person, with captains and the harbormaster and other concerned citizens, official and civilian alike. By Friday afternoon, the office on the third floor of the Levy Building had become a scene of constantly ringing phones and people crowding in to get their questions answered.

Joseph was overseeing most of the calls and other conversation. Isaac spent time on the roof, taking readings.

Then, as the afternoon waned, Joseph put John Blagden in charge of the phones and the people jamming the office. Joseph too climbed to the roof. Isaac went home, and Joseph relieved him.

Meanwhile the clouds had thickened. The day that had started clear was now cloudy. From out in the gulf, the swells kept coming.

Annie McCullough had recently bought a new pair of shoes. They weren't quite right, though, and that Friday she'd sent her husband Ed to the store to exchange them.

Living level with the beach, really almost on it, Annie could easily see and marvel at the huge breakers rising. She watched the people leaping about in the unaccustomed surf. She was thinking about the coming storm, about what it might do to the roses she grew in her side yard.

A neighbor rushed by. She called to Annie that she was heading down to the beach to have some fun.

But Annie didn't want to frolic. If the storm came, those roses might get battered.

Soon Ed returned from the shoe store, and Annie got Ed's thirteen-year-old cousin Henry, who lived with them, to dig up the roses. He put them, with their rootballs, in a big tub of soil for safekeeping.

The red-haired young schoolteacher stood where the street ended at the gulf beach. It was late Friday afternoon, and Daisy looked out over the expanse of sand and water. She watched each huge wave rise, then tower and fall, pounding the sand. Between long pauses, she heard the roar of the swells.

She saw clouds everywhere in the great gulf sky. They had grown so thick that they obscured the sunset.

Daisy decided that this Friday evening would not be an ideal evening for her usual daily swim. It had been a nice morning. It wasn't a pleasant evening.

Before turning for home, though, she watched a moment longer. The gigantic waves, as they rose and crashed, looked dull brownish—full of sand. The bathers drawn earlier by the novelty of surf in the gulf had all gone home. Nobody was in the water now, only a few observers like her, standing here and there about the beach. Watching the waves.

So Daisy went home to the apartment where she lived with her mother, aunt, brother, and sister, in the Lucas Terrace building, at the far east end of Broadway, near the beach.

Because it was a normal Friday evening, Rabbi Henry Cohen would have been leading services at Galveston's Congregation B'nai Israel and planning the next day's Shabbat services. But

Rabbi Cohen's services were something special—as was the man himself.

Cohen had begun making some radical changes in traditional practice. Back in Woodville, Mississippi, where he'd served before coming to Galveston, he'd noticed a problem that was disadvantaging the Jewish farmers. On Saturdays, farmers would cart their produce into town, set up stalls, and sell. Observant Jewish farmers, though, couldn't open their stands until after Shabbat services were over. All morning, they could not compete.

So Cohen began ending services early, unleashing Jewish farmers into the market fray. People took notice, but instead of sparking conflict, Cohen's decision made him new friends. Public speaking was the major form of popular entertainment; people liked all kinds of lectures, sermons, and recitations. When Gentile farmers got word that the rabbi was preaching a powerful sermon on Saturday mornings in Woodville, they started showing up at the synagogue to listen. So popular did Cohen's services become with all of the farmers that the market stayed closed on Saturday mornings until Shabbat services ended.

Rabbi Cohen had continued his unusually ecumenical practices here in Galveston. He became fast friends with Father James Kirwin, the Roman Catholic rector of Galveston's splendid, gigantic St. Mary's Cathedral. The rabbi was known and respected throughout the city, and he was seen everywhere, scooting about on his bicycle with a long list of good works to perform for people of every race, creed, and color. He was an odd duck—an Englishman, for one thing, with a slight stammer—and yet both his preaching and his endlessly energetic charity made Henry Cohen one of the favorites in town.

He read ten languages. He was a published Talmudic

scholar and historian of Texas. After studying at Jews' Hospital and Jews' College in London, he'd lived in Cape Town, South Africa, and served as a rabbi in Kingston, Jamaica. He became the Woodville, Mississippi, rabbi when he was only twenty-two, and Galveston's B'nai Israel rabbi when he was twenty-five. With his wife, Mollie, he had two children.

This English rabbi had visions for Galveston, and for Texas in general. One of them was to make Texas a major port of entry for immigrants: not only Jews but Catholics as well. The big wave of immigration from eastern and southern Europe had begun after the Civil War. Cohen wanted to help integrate the new Jewish immigrants who spoke Yiddish with the established German Jews of Texas. The northeastern port cities were teeming. Cohen wanted Jews and other immigrants to settle in the country's midsection and West, from Texas and New Orleans to the Rockies.

Around town, the rabbi's personal efforts were legendary. As he buzzed around on his bicycle, long coattails flapping behind him, he consulted his cuffs, where he kept his list of appointments. Like all of his fellow Galvestonians, he had plans.

Friday evening, Ed Ketchum, Galveston's genial, Yankee-transplant chief of police, was having supper at home with his family. Ketchum's job naturally made him more alert than other Galvestonians to changes in the weather that might affect public safety. All week, as he'd been catching up with paperwork after his trip to Chicago, the chief had been keeping an ear open to Isaac Cline's reports.

Earlier that week he'd received word, he now told his family over supper, of a bad storm over Cuba. And that storm had been moving north.

Also, earlier today, the Galveston weather station had reported rising wind. But those gusts were really pretty light, Ed assured his family, and the coming storm probably wouldn't amount to much.

All day, the Cline brothers had been fending off confusion and worry. Their feelings were based on the suddenness of the change, the reporting from Washington, and the direction of the storm. Now, Friday night, on the roof of the Levy Building, Joseph Cline began succumbing to a sense of impending disaster.

New Orleans was the nearest weather station to the center of the storm. The reports Joseph was now receiving had it that the storm was southwest of New Orleans and moving westward.

Joseph knew what that meant. The storm was heading straight for Galveston Island and the Texas Gulf Coast.

He didn't panic. He did his job. Down in the office, he quickly created a new weather map based on the reports he was receiving by cable. He left the building and took the map to the post office. He deposited it there to await the first train over the railroad bridge to the Texas mainland.

That was about midnight. Out of those thickening clouds, rain had started falling steadily. Fretful, Joseph went to Isaac's house, where he too lived, at Twenty-Fifth and Q, near the gulf beach. In his room, Joseph tried to sleep.

And he did sleep—but restlessly. Visions of hurricanes kept invading his dreams.

At 4:00 A.M. he awakened with a start. He had a sudden, clear impression that gulf water had flowed all the way into the Clines' yard.

Joseph got up. From a south window, he peered down at the yard.

It wasn't a dream. The yard really was underwater. The gulf was in town. Joseph went to awaken Isaac.

The secretary of the Galveston Cotton Exchange, Dr. Samuel Young, was an amateur meteorologist, and he lived near the beach too, only a block north of the Cline home. Sometimes during nighttime thunderstorms, from his upper porch Young could see Isaac Cline out on his own upper porch, observing the weather.

This Friday evening, Young had walked over to the beach to observe the unusual conditions of the waves, passing the Clines' on his way. Despite the normal storm-warning flag, Young had an idea that something worse was coming.

His concern was based on a visit he'd made to the Weather Bureau office in the Levy Building earlier that week, both as an official of the Cotton Exchange and as a meteorology enthusiast. Young had long wanted to see for himself how the mapmaker there drew the maps that his office received every day from the bureau's weather station in Galveston. The mapmaker used a system of colored chalk on a blackboard. To those in the know, every detail could be gathered for much of the country: tiny circles, letters, arrows, and dotted lines indicating rain, wind, air pressure, temperature, clouds, clear skies.

Young watched closely that morning when the mapmaker drew the symbols for the storm that had been observed at Key West. Young thought conditions there indicated the presence of a serious tropical storm—a cyclonic structure formed around a discernible eye, growing in strength somewhere south of Florida. Tropical storms like that can quickly turn into hurricanes.

The mapmaker, though, said he'd heard nothing from Washington about anything like that. Anyway, there was no chalk symbol for such a storm.

All week, Young had gone on thinking about it. All week, he'd thought the tide was unusually high, given that the wind wasn't behind it.

So as he stood on the beach Friday night, he was developing a theory. When he looked southward at the surf, he could feel the wind at his back. A north wind like that should be flattening the gulf, pushing the tide back. Yet the surf kept rising to crashing heights he'd never seen before. On that basis, and despite the bureau's refusal to call this storm what it was, Young concluded on his own that this thing approaching the Gulf Coast was a hurricane.

His wife and children were on a train, coming home to Texas from a summer in the West. Because hurricanes on northwest paths always, the science had it, recurve eastward. Young concluded that its center was likely to hit not Galveston but Mississippi. He left the wild beachfront and went home to bed.

At 4:00 on Saturday morning at the Cline home, Isaac's wife, Cora, and the three little girls slumbered on. Cora was pregnant again.

But the brothers were wide awake. The worst had begun, Joseph told Isaac. Their yard was underwater. Joseph was sure that what was coming was a disastrous hurricane.

CHAPTER 8

SATURDAY MORNING: STORM TIDE

Isaac Cline remained as methodical as ever.

The station had a regularly scheduled 7:00 A.M. report to make to Washington. Conferring quietly with Joseph in the early hours of Saturday morning, Isaac decided that before sending that report, the brothers must be sure to assemble all data, both from weather-station instrument readings and from their own observations of conditions. They must then cable updates to Washington every two hours as matters in Galveston developed. He sent Joseph back to the Levy Building to take readings.

By 5:00 A.M. Isaac himself had harnessed his horse to a two-wheel hunting wagon. That would keep him from having to wade through water that was already covering the beaches.

He drove the few blocks to the gulf side. Then he turned

east and drove all the way out to the east end, not far from where Daisy Thorne lived with her family at the Lucas Terrace Apartments.

There the beach was wide, and the water had farther to come before it reached the streets. Looking out over the expanse of sand, Cline now began timing the swells.

They came every one to five minutes. They were increasing in force and frequency. Even here, on the broad east beach, the water was moving steadily up the sand toward the street.

Sometimes he sat in his horse-drawn wagon in the rain. At other times he paced on the beach. Always he kept looking at the water flowing from the gulf toward the streets.

And now Isaac Cline made another observation.

This one was decisive. And it was bad.

Along with the state-of-the-art instruments at the Levy Building, Galveston had a state-of-the-art mareograph—a tide gauge. As early as the 1850s, coastal U.S. cities had been using self-registering gauges to mark the heights and depths of their tides.

Now, on Saturday morning, the tide gauge told Isaac Cline that something was very wrong.

The gauge was essentially a metal tube with open ends, fixed in the water near the shore to still the effect of waves and splashes. The tube was therefore known as a "stilling well." A float in the tube rose up with the flow of the tide, down with the ebb.

The float was attached by a wire-and-pulley system to a pen sitting on a paper drum that rotated steadily above the tube via clockwork. Moving in response to the height of the float in the water, the pen etched the water level on the paper on the drum. The result was a graph of highs and lows over time.

This Saturday morning in Galveston, the tide gauge showed the water level at 4.5 feet higher than normal. Which would have made sense—if the wind were coming from the south or southeast, from Cuba, that is, across the gulf itself, thereby pushing water from behind up the beach and into town. That wind direction would easily explain a tide so high.

But the winds this morning were not behind the water. The wind was out of the north. It pushed against the tide. And it was fairly stiff, at 15 to 17 miles per hour.

So the gulf tide was rising to an extraordinary height *against* the wind. That never happened.

Not that the north wind was failing to suppress the waves and hold back the tide. That was the scary thing. It *was* holding back the tide. And still the tide was almost five feet above normal.

This was just the phenomenon that had been bugging Dr. Young of the Cotton Exchange. Now Isaac saw it too. High tide, with wildly crashing surf, but driving against the wind, and not pushed by it: this was a storm tide. It meant serious trouble for Galveston.

Meanwhile, gulf water was moving slowly, steadily from the beach into town. It wasn't only the Clines' yard that was underwater. In the main part of downtown, on the gulf side, nothing stood between the sand and the first rows of houses—no breakwater, no seawall, no barrier of any kind—and there wasn't even any real height for the water to climb. It just flowed straight into the streets. The beach was nearly level, and the streets flowed levelly from the beach and into town.

So now gulf water had not only inundated the beach at the bathing pier. It began moving into some deep channels, formerly known as Galveston's city streets.

And yet just as on the day before, some Galvestonians' first impression of the arrival of the water that morning was sheer fun and excitement. Despite the steady rain and the rising flood, all Saturday morning people waded and splashed through the water down to the beach to marvel anew at the gigantic, crashing surf and the water flowing in the streets. All morning, and even into the afternoon, as the rain pounded and wind began blowing with supernatural force, and the sea rose, Galvestonians played in the water.

In addition to her love of bicycling, Daisy Thorne was a photography enthusiast, and Saturday morning she was up early to take pictures of the towering surf. She stepped onto the porch of the two-story apartment at Lucas Terrace, which stood conveniently near the broad east-end beach, now going underwater, and pointed her camera at the gulf.

The pneumatic tires on Daisy's bicycle represented an improvement, so she'd been early to adopt them. By contrast, the new film cameras—such as the point-and-shoot Brownie introduced that year by the Eastman Company—really were meant for snapshots, not a true advancement in photographic quality. Daisy was an artist. She hadn't adopted film; she stuck with professional-level gelatin dry plates.

After inserting each emulsion-prepared glass plate into the big camera, she waited seconds on each exposure, hoping to capture the drama of the glowering clouds and spraying surf.

People were wading down Broadway, she saw. They were coming to the east beach in the rain to get a look at the astonishing scene and hear the big roars.

By 8:00 that morning, the water nearly surrounded Daisy's apartment building. By then, Daisy had used up all of her plates. She had some embroidering to do, so by midmorning, she was inside, working on the embroidery—it was a pillow for her fiancé Joe Gilbert's new doctor's office—and considering the weather.

This was a wild and flooding storm, Daisy knew. But though built near the sandy beach, the Lucas Terrace building was as solid as a rock. The water kept creeping steadily around it.

The Cline brothers must have already been horribly aware on Saturday morning that it was probably too late to avoid an awful disaster in Galveston. Their subsequent memories would by no means agree perfectly—yet in later years, both brothers recalled making strenuous efforts that Saturday to give warnings and save lives.

Isaac Cline had long believed in two dovetailed theories. One was that hurricanes could not travel from Cuba to Texas. The other was that, when bad gulf storms did make landfall on Galveston, the bay allowed release for all that destructive energy, ushering it onto the low marshes of the Texas mainland and leaving the island unharmed.

So it's hard to know exactly how Isaac read the situation in Galveston as early as Saturday morning. He seems to have been struck by a responsibility to warn as many people as possible of the likely force of the coming storm. According to his later recollection, Isaac began driving his two-wheel wagon among the beachfront houses. He personally advised people vacationing there to leave the island right away, as quickly as they could.

He spoke to people living farther back from the beach, three blocks or so from the water. Their houses' foundations, Isaac told them, could be knocked around by a strong ebb and flow of tide. The tides might bring their houses down. He advised people to seek the high ground.

The "high ground": despite Galveston's sandbar-like position between the gulf and the bay, that term did have some meaning. People knew from long storm and flood experience where the high ground lay.

A spine runs through the island on the east-west axis. It tapers down to sea level at the east end; well out of town to the west, it breaks up into bayous, along with the rest of the island.

In the main part of Galveston in 1900, the high ground was maybe eight feet above sea level. Land sloped downward from the flattish top of the spine: southerly, toward the water of the gulf, and northerly, toward the water of the bay. The spine was called Avenue J—otherwise known as Broadway.

There, where the astonishing mansions of the Gilded Age millionaires lined the wide avenue parallel to the gulf beach and the bay harbor, lay the city's highest ground. It didn't run through the exact middle of town; it lay closer to the bay than to the gulf. Whenever the low streets on the gulf side had gone underwater, in living memory Broadway had always remained dry.

So it's not surprising that the Broadway area is where Isaac Cline would have sent sea-level citizens on both of the water-exposed sides of town to seek shelter. In this storm, however, the high ground on Broadway would soon cause a condition that neither Cline nor anyone else in Galveston could have foreseen. High ground would offer no help, putting it mildly, in withstanding this hurricane.

Just outside the family rooming house, Louise Bristol was playing in the water with her friend Martha. The weekend had finally come, with relief from the excitement and nervousness of that first week of school. And the weekend was turning out even better than Louise could have imagined.

There was crashing surf, and relief from the heat, and now the curious delight of brown water filling the deep streets. The water slowly changed the landscape before their eyes into a magical place the children had never seen before.

Louise was thrilled in particular by this idea: she didn't have to go to the beach. The beach was coming to her.

Her mother, Cassie, by contrast, was not having fun. But it wasn't the water in the streets—that didn't concern Cassie much. She was just busy today, like every other day. As the landlady coped with the impact on her household of the imminent return of medical students who roomed at the Bristols' home, Cassie couldn't be bothered with worrying about rain and wet streets.

So Louise and Martha splashed, jumped, and swam. They played in what had once been their street and was now a deep, strongly flowing river of salt water.

Arnold Wolfram was getting ready to go to work, and members of his family were not among the cavalier Galvestonians. They pleaded with Wolfram to stay home, safe from the raging storm.

The grocery salesman had a past in the wild Texas West. He'd hung out with Texas Rangers and famous badmen. He reassured his family: all would be well.

As soon as things got bad, he promised, he'd head home. They weren't reassured, but Arnold went off to work anyway.

By late morning, people were still cavorting in the gulf breakers. Drenching rain slammed gawkers' backs as they gazed southward on the beach at the gulf. But the wind was blowing so powerfully out of the north, against the tide, that their fronts stayed dry as they watched the towering surf and the crazy people playing in it. Even as people watched, the midway shops and stands near the bathing pier started collapsing into the wind and water.

Not everybody was so relaxed. Annie McCullough had her rosebushes safely in tubs. Yet overnight she'd felt the winds pick up, and now the rain was falling hard. The beach was underwater, and Annie's street was becoming invisible, one with the gulf. It wasn't just the roses that seemed vulnerable now.

Annie and Ed decided it was time to go. They were ready to evacuate their house, so close to the beach; they would seek shelter up on the Broadway high ground until the storm passed.

Annie didn't feel frightened. She was just being sensible. Ed hitched his mule to his dray, the low delivery wagon with no sides and two big wheels. He and Annie and young cousin Henry drove the mule through the water to pick up Annie's mother, who lived nearby. Annie's father, the government customs man Fleming Smizer, was off doing his job at Point Bolivar on the mainland. They expected to see him soon.

Annie's mother clambered up on the dray, as did a number of other relatives, young and old, and Ed hoisted up a trunk full of clothes that Annie's mother had brought. The plan was to head to the big, strong, brick-and-stone school building near Broadway, on Eleventh Street.

Just seven blocks uptown from the McCulloughs' house, the school building took up the entire block between G and H Streets. The building was only twelve years old, with the chateau-like spires typical of a certain grand style of 1880s public architecture. It had three gigantic main floors, plus high attics, and everybody knew that the Broadway area offered the highest ground in Galveston: that street had never flooded. The school looked like the safest convenient place to wait out what was starting to feel like an unusually powerful storm.

The dray was crowded, so Annie said she'd walk the few blocks. Ed snapped the reins and the rest of the family rode uptown toward Broadway.

Rabbi Cohen had just finished leading Saturday services at the ornate, Gothic-revival synagogue of Temple B'nai Israel, a few blocks from Broadway toward the gulf side of town. He was heading for his house on Broadway in the rain on foot—not on his Cleveland bicycle, as this was the Sabbath—and that's when he saw all the people moving uptown through the storm.

Scores of them. All races, all ethnicities. Just like the Mc-Cullough and Smizer families, they were moving to the higher ground around Broadway

Whether in response to warnings by the Clines to people living along the gulf side, or simply realizing it would be smarter to get moving, a long line of people was flowing up Broadway from the east end of town—reversing the flow Daisy Thorne had seen earlier, when people had been drawn down Broadway to the east beach to goggle at the surf. The people were carrying suitcases, trunks, boxes, and household items, even lamps, even photographs.

These residents were serious about getting out of the way of the storm. Rabbi Cohen also detected a holiday mood. As the crowd slogged uptown through the driving rain, boys ran ahead to take long slides in the mud of Broadway's median esplanade.

It was characteristic of Cohen to start figuring out how he could help these refugees. Some of the people told him something they'd seen down at the water. The trolley trestle that ran over the water just off the beach was underwater now. It would soon collapse. The entire bathing pier and its midway shops and stalls were wrecked, they told Cohen. The striped tentlike bathhouses were going into the water.

Cohen ran into his Broadway house, rustled up umbrellas and blankets, brought them back outside, and began handing them out to the people on the moving line. His wife, Mollie, came out into the driving rain with a bushel of apples, and as the people passed them, the Cohens handed out the apples too. Soon the rabbi was drenched. Mollie finally got him to come inside and change his clothes.

When the Cohen family sat down to lunch, wind was howling around the house. It was dark. Mollie and the rabbi began trying to reassure their children. They lit candles, and Mollie recalled the storm of 1886.

Her father's Market Street store had been flooded in that storm. But water had never come all the way up to Broadway.

Just then the entire house rocked once. Plaster fell from the ceiling and crashed to the floor. "Just a little blow," Mollie reassured the children.

Meanwhile, out where the Gulf of Mexico meets Galveston Bay, off the northeast end of the island, in the passage between the island and the Bolivar Peninsula, any difference between bay water and gulf water was becoming meaningless. The rising, crashing water of the gulf began pouring quickly through that channel into the bay. The bay was rising too, and it was being attacked by wind.

Ships were moored in that channel. Their captains began trying to cope with the churning water that rocked their boats.

Captains of steamships fired up their boilers. Having some control over their crafts might help them withstand the growing chaos.

In the Weather Bureau office in the Levy Building, phones kept jangling. People were calling for information—they simply dialed 214 on a rotary phone—and the office was even more jammed now with worried people than it had been the day before.

Not everybody was splashing about and having a party. Not every Galvestonian was cavalier about weather. Many were getting nervous.

In the later memories of some citizens, they were given no real warnings by the Weather Bureau office of how bad things were about to get. Some would later recall having their fears pooh-poohed.

In Joseph Cline's future recollections, he and Blagden, coping with the pandemonium in the office, advised people to move to the high ground. They advised vacationers to take trains off the island. Soon they'd have to stop giving that advice; soon trains would stop running.

At 11:00 A.M., Joseph took readings. The barometer was falling fast. Winds were thirty miles per hour—still out of the north. Joseph sent that news via telegraph to Director Moore in Washington.

On Twelfth Street, a riderless horse came cantering through the wild rain, spooked by the storm.

A front gate had been left open by people going down to stare at the water on the beach. The horse turned at that open gate and entered the yard.

The horse climbed the porch steps. The front door was open too. The horse nosed its way into the house. It went straight up the staircase to the second floor.

Out on the east end, Daisy Thorne and her mother found themselves welcoming refugee neighbors. Many were fleeing their fragile wooden homes. The Lucas Terrace apartment building seemed far more solid.

The beach, viewed from the Thornes' windows, wasn't a beach any more. It was gulf.

On their wood stove, Daisy's mother made coffee. She and Daisy started cooking biscuits for the visitors. By midmorning, Lucas Terrace was standing in the water. The gulf surrounded it on all sides, and rain whipped the windows. Daisy and Mrs. Thorne kept welcoming guests. The place was getting crowded.

Across town, another big, solid building was also surrounded by water. This one was near the beach on the west side: St. Mary's Orphanage.

Early that morning, Sister Elizabeth Ryan had left the orphanage to buy supplies at the markets in town. By 10:00 A.M., the water was three feet deep on all sides of the building, and Sister Elizabeth hadn't returned. All morning, as the weather had grown more violent, Mother Camillus Tracy had grown more worried. Finally, she'd sent two of the orphanage's maintenance men to town to track down Sister Elizabeth.

Now it was midday. The water surrounding the building was higher still. None of the three had yet returned.

Louise and her friend Martha had been enjoying splashing in the deep water in the streets. There were tiny toads leaping out of the water—even some snakes going by.

But now they were surprised. Boxes were floating quickly by them on the water. Then some clothes. Some loose boards. Some toys.

These objects were bewildering. The fun seemed to be ending. The girls said good-bye and splashed homeward.

Only minutes later, Louise was watching from a chair in the rooming house kitchen as brown water flooded straight across her mother's carefully tended gardens.

Police Chief Ketchum, working at his desk at City Hall, was having as busy a morning as the Cline brothers and their assistants. While Joseph Cline manned the Weather Bureau office in the Levy Building, Chief Ketchum was trying to cope with the rising anxiety of dozens of Galvestonians.

His phone rang steadily, and by late morning, he wasn't hearing only from people with questions and concerns about the storm. Now he was being asked to make rescues.

Galveston had a horse-drawn police patrol wagon. Ketchum started sending it out to help people who had been stranded in their homes by the rising water.

One of the first patrol-wagon missions went to Daisy Thorne's building, Lucas Terrace. The chief had a call from a man whose mother-in-law lived in one of the apartments there—on the third floor, just upstairs from Daisy and her family. The mother-in-law was stranded, the man said. People in the building had been calling on the phone for hacks—horse-drawn taxis—but no cabby was willing to fight the storm to go that far east. The building was already surrounded by water.

So Chief Ketchum dispatched Officer J. T. Rowan to drive the patrol wagon out to Lucas Terrace. Rowan's mission was to evacuate the mother-in-law.

Officer Rowan, pushing his team through driving rain, arrived on the east end to find that water was indeed rising quickly around the apartment building. He slogged in as quickly as he could.

Once inside, and with the mother-in-law safely in tow, he knocked on other doors, including Daisy's, to check on residents. Officer Rowan suggested to Daisy's mother that the Thorne family take the ride uptown in the patrol wagon. Why not grab this chance to get off the flooded beach and seek safety?

But Mrs. Thorne protested that she had a house full of guests, all the neighbors who had abandoned their frame houses and come to Lucas Terrace. She and Daisy were in the midst of making them biscuits. And the building was safe. That's why everybody had come there.

"These people will be hungry," she told Officer Rowan. As Rowan drove the man's mother-in-law through wind and rain back to town, Mrs. Thorne, Daisy, and the rest of the family stayed in the apartment in the building in the water.

Some people were still taking the characteristically cavalier Galveston attitude in the face of wild weather. On a trolley car at lunchtime, riders didn't even discuss the storm raging on the street. Then, at about 12:30, the trolleys stopped running.

At Ritter's Saloon too, the businessmen gobbled their lunches in the rowdy, smoke-filled room, making deals, trading jibes, and cracking jokes about the weather.

And yet vacationers had spent the morning trying to flee the island. They'd been mobbing four-story, red-brick Union Station on the Strand near the bay, eager to pack onto the morning trains that crossed to the mainland on the railroad causeway. One of these trains, everybody now knew, would be the last to get off the island until the storm had passed. Meanwhile, at the harbor, captains of every kind of craft were battening down the hatches and crossing their fingers.

Still, on some wharves, men continued loading cargo. Even as late as early Friday afternoon, a solitary group was working on a wharf. They were trying to load bags of flour onto a schooner whose prow had begun riding with each wave as high as the tops of the portside buildings.

The ship, the flour, and workers were all soaking wet. Yet the men kept working.

CHAPTER 9

SATURDAY AFTERNOON: "HALF THE CITY UNDERWATER"

DAISY WAS LOOKING OUT THE WINDOW, AWAY FROM THE beach, at the wind sweeping rain along Broadway and the water rising above the street. The officer with the rescue wagon had left Lucas Terrace. The Broadway telephone poles, rocking in the wind, had dropped their phone lines into the water. The poles themselves wouldn't stand long.

She watched a cottage across the street—it belonged to one of the neighbors eating biscuits in her apartment—start succumbing to the raging storm. It didn't take long: within seconds, the house was down and gone.

Soon the water would be high enough to pick up that shattered house and carry its timbers, along with the timbers of hundreds of other houses, turning them all into massive tonnage, a heavy tide of debris moved by the swirling flood.

Behind Daisy, as she stood at the window, the Thorne apartment was now jammed. Some were in varying states of loud consternation. Some were calmly eating Mrs. Thorne's food. Some hid their eyes to avoid seeing the destruction going on outside.

Downtown near the beach, the young lawyer Clarence Howth was wading up the street to his big, solid, raised-up home with upper and lower porches, three blocks from the gulf. He'd just had his midday meal at Ritter's, as he so often did. Then he'd tried to drive his wagon home.

Encountering water three feet deep in the street, Howth hitched the horse and wagon to a post. He proceeded on foot.

Now he was splashing through gulf water. In some places, it came up to his shoulders. Howth was a man who normally showed magnificent unconcern. All that was about to change.

Gaining his front steps, Howth hauled his sodden self up onto his porch, removed his shoes, went inside, and climbed the stairs to the third floor to look in on his wife and new baby, the couple's first. The baby had been doing fine, but young Marie Howth was often unwell; her pregnancy and delivery had been a strain, and she was confined to her bed.

Howth went to his bedroom, disrobed, dried off, and changed clothes. He returned to his parlor.

The builder had planned for floods. This house had a raised ground floor; it sat on a raised lot. Howth pulled up a chair and sat at a parlor window facing the street. Leaning back to watch the amazing show outside, he lit a cigar.

Arnold Wolfram, to the great relief of his wife and children, did come home for lunch as he'd promised.

But after lunch, he told them he was going back downtown to work. Both his wife and their children started crying.

Arnold was a stubborn man. He was not to be deterred from his course by tearful pleadings. He headed back into the storm to do his job.

Meanwhile, at City Hall, Chief Ketchum was realizing he'd be here a while. He'd gotten wet, so he called home and asked his wife to send over a change of clothes. Mrs. Ketchum quickly put together a package—heavy pants, a flannel shirt, boots—and the Ketchums' son Henry drove a buggy through the rain to get the clothing to City Hall.

The chief was meanwhile ordering the patrol wagon back out to the east side. Not as far as the Thorne apartment at Lucas Terrace—nobody could get that far now—but to a home closer by, where a family was evidently stranded.

When those officers returned to the station, they brought with them nearly a hundred wet, upset people, all crowding into City Hall. The officers reported that when they'd arrived at the scene, they'd been unable to cross the water to the house. So they'd requisitioned a boat. On the boat, they evacuated not just one family but all of these other people from that entire part of town.

With City Hall jammed now, some of the people were getting hysterical. Henry Ketchum arrived, soaked from his trip, and gave his father the dry clothes, but then, because the scene in the police station was so intense, he stuck around.

After a while the chief realized his son was still there. He sent Henry straight home. But he said nothing more than "It's going to be pretty rough tonight."

Out on the street, Henry got the message. As he tried to make his way home with the buggy, slate roof tiles were flying through the air; he had to duck as he drove. He was on Thirty-Fifth Street, a fairly high street. Yet the water was up to his wagon's axles.

A barn that the Ketchum family owned stood a half block from their house. There Henry left the horse in its stable and walked out into swirling water, now up to his chest. He was beginning to fear for his life when he saw a Scottish terrier standing on top of a doghouse to which it was chained. The doghouse had bobbed and floated into a fence, where it stuck. The dog barked to get Henry's attention.

But Henry was scared now, not sure he was going to make it home. He kept pushing through the water. He left the stranded, barking terrier behind.

A train had left Houston for Galveston at 9:45 that morning. Around 1:00 P.M., it finally pulled into Galveston's Union Station on the Strand. Clearly its passengers were moving in the wrong direction—onto, not off, the island. But leaving Houston that morning, they hadn't realized what they'd be getting into.

Now they knew. Filled with trepidation, they'd made it through the rattling wind all the way across the bay railroad causeway through rising water. They'd gained the island.

But traveling on the island, their train had come to a full stop. The track was washed out.

Stranded in the cars, the passengers could only sit and wait. They watched the storm rage on the flatlands through windows beaten by rain. They watched the water rise inexorably until it covered the railroad tracks. Then it climbed the wheels. How high would it go?

After an hour or more of tense waiting, a new train arrived on a nearby track that was still intact. The stuck train started backing up. It backed half a mile to higher ground. There the passengers changed trains in the rain and wind, moving as quickly as they could from track to track.

The new train proceeded toward the city of Galveston—but it traveled at pedestrian speed. Literally: the train crew was wading through the water ahead of the engine. They were moving debris out of the way.

For the water wasn't just rising now: it was moving, streaming, flowing, flooding. It seemed to have a fast current, running hard from east to west. And it was carrying tons of wreckage.

Now the train passengers had arrived at their destination. But just to get to the station platform, the men had to stand waist high in the water and help the women and children disembark.

There were still some hacks waiting at the station to haul the passengers through the floodwaters, pulled by horses belly-deep, to hotels and homes. But hotels and homes didn't seem so solid at that moment. Fearful of venturing out, fearful of the water rising so quickly on the station's ground floor, they went up the station's stairs to a bare room on the second floor.

They hauled their baggage up those darkened stairs and huddled up there. They were cold and wet yet determined to wait out a storm that, they still believed, would no doubt be memorable, but would probably be fine in the end.

Then, below the stairs, they saw a body float into the first floor of the station.

A dead child.

A tourist from Scranton, Pennsylvania, was standing on one of Galveston's high sidewalks near the harbor and watching the stream rise to cover his feet. When the deep streets were full, the sidewalks themselves began to join them underwater. The tourist was beginning to get nervous. When he looked uptown, toward the wharves, he saw ships' prows weirdly high—competing with the heights of warehouses. Big horse turds rushed by in the river that had so recently been a street.

He saw a wading man slip in the water and fall. The tourist watched as the fallen man was borne quickly away, through the stream, laughing all the way.

Dr. Young, the amateur meteorologist and head of the Cotton Exchange, had realized earlier in the day that the U.S. Weather Bureau had placed the storm on the wrong track. Earlier that week, Young had already questioned the complacency of the Weather Bureau regarding the Cuban storm. Friday night, he'd noted the bizarrely high tides rising inexorably against a high north wind that should have been holding them back. This morning he'd gone out again to look at the beach, and then he knew this was a hurricane.

Immediately he walked up to the Strand, entered the Western Union Office, and telegraphed his wife, who was with the children aboard a Southern Pacific sleeper on their way home. This cable, he knew, would find her during a scheduled stop in San Antonio. It warned her to stay in San Antonio until she heard from him. A great storm, Young told his wife, was arriving in Galveston.

Now, a little after 1:00 in the afternoon, Young was wading to his house near Isaac Cline's, just a few blocks from the gulf. He could take satisfaction, at least, in having kept his family from riding into the storm. And like Clarence Howth's, Young's house—indeed his whole property—had been set up to resist bad storms. The entire lot was raised five feet above sea level; even the front sidewalk was four feet up, with a deep curb into the street. The house sat on brick pillars rising four feet high.

And yet all over Young's stormproof lot was gulf water—at least a foot deep already. That meant the sea was at least six feet above its normal level on the gulf side of town. And the wind was still out of the north, holding the tide back.

Young sat alone on his first-floor porch and watched the water rise until the entire bottom porch step was underwater. But he knew how high his house sat. While he understood this was a hurricane, and was glad his family wasn't riding into it, he still had no fear that any water would actually come into his house.

At 2:30, Joseph Cline went up to fight the howling wind on the roof of the Levy Building and take readings. How much rain had fallen was now impossible to determine: the rain gauge had left the building, blown into the sky by the howling wind.

But the barometer was still falling, signaling that the worst was still to come. Joseph returned to the frantic office below. The Levy Building was well positioned to withstand a storm.

But now he received a phone call that confirmed his darkest fears.

Mollie Cohen had swept up the fallen plaster that the blast of wind had shaken loose. And yet outside Mollie and the rabbi's darkened home on Broadway, the wind kept roaring. The house continued to shake in the blasts. Plaster kept falling.

Rabbi Cohen pushed open his front door. Broadway was underwater.

This was the high ground. This had never, everyone believed, happened before.

The line of drenched people was still slogging along the Broadway sidewalks through the flood, carrying their treasured belongings. But any high ground remaining in Galveston was quickly disappearing.

Cohen watched the water rise over the bottom step to his porch. He closed the door quickly so the children wouldn't see.

Mollie began playing the piano: songs from *Patience* by Gilbert and Sullivan. The Cohen family sat in their dark house and sang.

After the unflappable young lawyer Clarence Howth left Ritter's Saloon and went home to smoke a cigar, other businessmen were still sitting around there eating oysters, puffing cigars, drinking beer out of big steins, trading jibes, and making deals. A storm couldn't scare them.

The wind rattled the front windows without stopping. They made jokes about it.

What they didn't know: that same driving wind had entered the building on the floor above, through open windows there. The upper floor was home to a print shop. Over their heads, the wind was pushing hard on the building's walls, bending them back and forth like paper.

The joists supporting the second-story floor couldn't stay put. At last the joists let go. The entire first-floor ceiling fell down into the restaurant.

With the ceiling came everything on the second story, sliding down and falling from above. That included the printing presses.

In Ritter's, two men were crushed in mid-joke. They died right away.

Three critically injured started dying more slowly on the saloon floor.

Others were only badly hurt: a waiter was quickly dispatched into the driving storm for a doctor. He was never seen again.

Annie McCullough had told Ed to take her mother in the dray, along with the rest of the family and their luggage. She could walk the few blocks to the school uptown, past Broadway. Annie still wasn't scared, just being sensible.

And yet as she slogged her way northward to the corner of Ninth and Broadway—only a block from the school at Tenth— she ran into a wind that pushed her backward. By now Broadway was not only full of water but also full of high waves, a streaming ocean.

Across that sea, toward the north side of the street, she caught a glimpse of Ed and the family on the flat wagon. But Annie stopped.

She couldn't cross. She was cut off. Now she was scared.

Louise was helping her mother and her older sister Lois, along with one of her older brothers, rapidly empty kitchen cabinets. After water had poured across her garden, Cassie Bristol had realized the danger the flood was about to cause her family and the entire house that was their living.

Now the garden was deep underwater, and Louise knew how hard her mother worked, how little time she had to do things like gardening. It upset Louise to think that her mother's work was all ruined.

Briskly, Cassie was now directing a decampment to the second floor with as much food and valuables as the family could carry. Louise, as little as she was, helped with the lighter items. Her eldest brother had not yet arrived—but now Louise saw him, through the window, wading slowly and with difficulty toward the house.

The water was so high now that he was holding his arms away from his sides, trying to steady himself against the swirling currents. The sight scared Louise.

Louise had a little Maltese kitten. It was acting skittish. It kept following Louise as she worked.

Then water started flowing in under the front door. It was almost time to get upstairs. But first her mother did something that startled Louise.

As the little girl watched, Cassie went into a shed attached to the house, where she normally chopped wood for the stove. She grabbed the axe and returned to the kitchen. She swung the axe over her head and began chopping holes in the kitchen floor.

Then she went into every room on the ground floor and chopped holes in those floors too. She was hoping to save her house by giving the water a controlled way in, easing the pressure from below. This way the rising water might not lift the house off the foundation.

The water was coming in under the door faster than it went down the holes. Its speed amazed little Louise. Her mother was still reaching down into the water to pull items out of the lower kitchen cabinets, but at last she had to stop: when bending, she couldn't keep her head out of the water. Finally the Bristol family retreated upstairs with everything they could move.

Annie McCullough crossed the ocean waves on Broadway the only way she could: carried in the arms of a big man who had stopped to help her. When the man had waded her across to the north side of the street and put her down, she saw Ed's dray floating now, the mule swimming.

Annie clambered onto the dray. The mule swam, pulling them. Everybody was soaked from the rain and the ocean. The men lay flat on their bellies on the wagon, holding the small children to keep the wind from blowing them into the water.

At last they arrived at the school. The water there was a bit lower. And yet as they left the horse and the dray and entered the building, the wind was so strong that it threw people about. The wind kept forcing the doors open, admitting gulf water in waves and rain water in sheets. Men were leaning against the doors, trying to hold them shut.

The rest of Ed's family—they lived near Annie and Ed—were already inside the building. When they saw Ed and Annie come in, they cheered. But upstairs, the school's second floor was a chaotic refugee scene.

The darkened rooms were packed with people. Many wept. Some wandered, calling out for their families. Some who couldn't find their families began crying out with grief and fear. Everybody knew now what was happening to Galveston.

Ed and Annie and their families went down to the first floor. They chose a place to sit in the east-west hall. The doors strained against the men holding them closed as Annie sat and waited.

All day, per Isaac Cline's instructions, the Galveston weather office had been sending reports to Director Willis Moore in Washington. The first morning cable from Galveston had reported Isaac's frightening observation of the nature of the storm tide: "Such high water with opposing winds," that cable said, "never observed previously."

But the last cable the Clines sent that afternoon conveyed a different sort of message. For it reflected a horrific reality.

At exactly 3:30 that afternoon, Isaac Cline—always punctilious about noting the time—finally recognized the situation for what it was. This was nothing less, he saw at last, than a deadly disaster.

By this time, the winds across the bay were gusting to 75 miles per hour, soon to become gale force. And that north wind had started to shift. It was moving from northwest to northeast.

So this, Cline knew, was indeed a hurricane. The gale wind was circling around an eye. Galveston was right in the hurricane's path.

The circling wind signaled other disastrous complications. For as the hurricane rotated around its eye, the wind direction too would circle the compass. It would become a wind out of the east, then a wind out of the east-southeast, then a wind out of the south-southeast. And finally the gale would blow straight out of the south.

While the wind had been coming from the north, it had been pushing against the gulf tides. The tides had nevertheless risen far beyond anything seen before. But as it circled all the way around and began blowing the gulf from behind, it would stop holding the water back. Worse: the wind would get behind the water and push it. There was no telling how high the water would then rise over the island.

Isaac was back on the beach when this revelation hit him. Joseph was at the Levy Building. Isaac hurriedly called the weather station from a phone near the gulf beach.

Joseph picked up the receiver. He listened closely amid the hectic crowds in the office, and his heart sank as Isaac dictated the wording of a new cable to Director Moore in Washington.

This cable wasn't just reporting. It was also forecasting.

There was about to be great loss of life, Cline dictated to Joseph. Galveston would need immediate relief.

"Gulf rising rapidly," the cable read. "Half the city underwater."

For it was true. Not only had the waters submerged the deep gutters, then the deep streets, and then the sidewalks themselves, but wind was driving the rain hard now, adding to the water's volume. And wind was knocking into the water and launching into the sky everything nailed down and otherwise. At the gulf, the tall, gaudy bathhouses at the bathing and amusement pier had swayed and toppled. They floated, they bobbed, they sank.

At the bay harbor, the waters had risen a full six feet, and waves were slamming the piers. Both steam and sailing ships' captains had tied their crafts to wharves and moorings with every bit of line they could find. They dropped anchors too. The steamers kept their boilers going. Anything to keep these rocking, leaping boats from disappearing into the chaos.

Down at the gulf, tracks for the trolley that ran along the island over the water had gone under. Up on the bay side stood far more important transportation infrastructure, the railroad bridges and the wagon bridge.

Earlier that day, a train packed with people had left Galveston's station, heading for Houston. Just like the unhappy passengers who had come into Galveston, the Houston-bound passengers crossing the bay could see the rising water getting close to the tracks. Now both the rail and the wagon bridges were wrecked. By the time Isaac dictated his forecast of lost life to Joseph for telegraphing to Moore, there was no way off the island.

But maybe information, at least, could make it off the island: that cable to Willis Moore with a plea for relief from the horror that the Cline brothers now had no doubt was about to ensue.

Joseph Cline stepped out of the Levy Building into the blinding rain and wind. Tucked away in his clothes was the encrypted cable with Galveston's desperate request for relief from Washington. He started for the Western Union office a few blocks away—normally a quick jaunt—to send the cable.

But the wood-block pavement of the business sidewalks on the bay side of town had come loose. The blocks were bobbing around in the high water like corks up at the level of the high sidewalks.

Joseph stepped into the water and waded knee-deep through the driving rain, making tortuous progress toward Western Union. The water was rising even as he walked.

Inside the Western Union building, the soaked meteorologist was informed that there would be no telegraphing there. The cable lines had succumbed to the wind two hours earlier. The Postal Telegraph Office might be an alternative

So Joseph waded there—yet another slow, drenching block. But no. Those lines were down too. Joseph fought his way all the way back to Levy Building.

Upstairs in the office, using the telephone, he asked the phone company operator for an immediate long-distance connection to the Western Union office in Houston. The operator refused.

There was only one operational phone line to Houston now, and it was working only intermittently. More than 4,000 calls were ahead of Joseph in the queue, the operator told him.

Joseph reminded the operator that according to protocol, rushed government calls had precedence over civilian calls. The operator still refused to place the call. Joseph hit the roof and demanded a supervisor.

The manager got on the phone, and thankfully, here was a man Joseph knew. He put Joseph through to Houston.

On the phone with the Western Union operator in Houston, Joseph read Isaac's message aloud. While reading it, he realized it was inaccurate. Half the city was not underwater. All of Galveston was underwater.

And yet Joseph also took pains to remind Western Union in Houston that the contents of this message were highly confidential. He was still concerned about the rivalry between Houston and Galveston. News of the submerging of Galveston, Joseph told the operator, was the property of the United States. Only Willis Moore in Washington could make it public.

Moments after that message made it to Houston by phone, Galveston's last phone line blew down. It joined the telegraph lines in the wild water. Silence prevailed between Galveston Island and the rest of the world.

The Clines' cable, however, did make it out of Houston.

In Washington, Willis Moore received it.

He wouldn't hear from Galveston again for days. By then, everything would be different.

Daisy's Thorne's mother had declined help from the police. Now the police could not have reached the Thorne apartment: the Lucas Terrace building was deep in what was now the gulf itself.

Water was ankle-high on the ground floor of the Thornes' two-story apartment. Yet Mrs. Thorne stayed in the kitchen there, wading to make biscuits and coffee for the terrified guests—dozens of them now, some of them already homeless— who filled the apartment.

Finally, though, Mrs. Thorne had to give in. She left her kitchen and went upstairs with Daisy and all the others.

But Daisy came back down to the first floor. The trick about a flood, people had often said, was to give it its head, relieve its pressure, diminish its power by letting it in. To that end, Daisy opened the doors on the first floor. In rushed the gulf waves.

The Thornes had five cats. In three trips, Daisy scooped them up out of the water and ran them upstairs. The water on the first floor was knee-deep and rising. The raging wind shook the building with every blast. The visitors were getting more and more terrified.

Daisy had begun today by photographing the waves. Now, as afternoon waned into evening, when she looked out the second-floor window at the dark and violent scene, she saw no other houses standing.

CHAPTER 10

THE NIGHT OF HORRORS

As night fell on Galveston, the storm only gained strength.

Isaac Cline's realization had been correct: the hurricane that was passing straight through the city was circling around an eye of drastic low pressure. This was the kind of storm that does not readily weaken easily but instead draws energy from a variety of sources, throwing its titanic violence in a multitude of directions all at once.

In 1900, what meteorologists feared most about hurricanes was the astonishing strength and wildness of the winds that accompany them. Hurricane-force winds rip roofs from buildings and tear up deep-rooted trees and structures on deep foundations. They toss those gigantic, heavy objects about as if they were nothing but balsa-wood slivers.

But the ripped-up debris is, of course, big and heavy. Objects become deadly missiles with bomb-like destructive power over people, buildings, everything. Smaller objects too, borne through the air with force—slate roof tiles, lighting fixtures, flowerpots, anything at all—add to this deadly barrage, which occurs on multiple trajectories, random and unpredictable.

Meanwhile, with the roofs shattered and the big timbers shivering and quaking, the wind robs buildings of all integrity, exposing them to the rising, shoving flood. From shacks to grand homes, churches, and public buildings: all go down, sometimes slowly and in pieces, sometimes all at once.

And the bigger the building, the more stone and brick involved in its construction, the deadlier to frail human life is its fall.

In 1900, meteorologists knew all that. But the winds produced by this cycling hurricane attacking Galveston were of higher velocity than those scientists believed was physically possible. The speed resulted in part from the air pressure at the deathly still center of this system, likewise lower than most scientists then believed possible. They thought air pressure could never fall as low as the pressure was in fact descending now on the evening of September 8, 1900, in Galveston, Texas.

Joseph Cline was shaken anew by the final barometer reading he took that day. Like Captain Halsey, at sea earlier in the week, Joseph was getting a reading below 29 inches. That was lower than barometers were generally known to fall.

So the Levy Building—an unusually solid structure—was actually rocking in the blasts of wind. And at 5:15 that afternoon, the wind gauge on the rooftop weather station was torn from its housing. It hurtled into the dark sky to join the rain gauge in oblivion.

When that happens, a wind gauge has done its work, in a crude fashion: it is reporting that the wind speed is terrifyingly high. The Galveston gauge's final official wind-velocity record-ing was 84 miles per hour for the previous five minutes. That period included two minutes at nearly 100 miles per hour. And the wind was getting higher.

And yet Willis Moore, the Weather Bureau chief, had always said winds could not reach those speeds, that such re-ports were anecdotal and hysterically exaggerated—the kind of thing those superstitious Cubans might come up with. To-night the Galveston storm was proving wrong both Moore himself and also many of the certainties on which he and the entire bureau based their practice.

Having barely gotten the final telegraph off to Washington, Joseph left John Blagden to man the office and started splash-ing toward the beach to give further warnings.

And at 7:15 P.M., Blagden, still alone at his post at the weather station in the rocking Levy Building, took a barome-ter reading that he could barely believe. The pressure stood at 28.48 inches.

Blagden would have reason to doubt the evidence of his equipment and his senses. That was the lowest official barom-eter reading that had ever been taken by any U.S. Weather Bureau office. And it was correct.

And yet these two record-breaking phenomena—both low pressure readings and wind velocities that until then were inconceivable—were connected in Galveston to another hur-ricane phenomenon, and meteorologists of the day were less sensitive to it than they were to wind and pressure. This was the rising of the gulf waters.

Weathermen of 1900 feared hurricanes mainly for the

winds, not so much for the floods. But the rising of the gulf on September 8 was causing a "trap" effect on Galveston Island. This phenomenon reflected a condition that precisely contradicted Isaac Cline's confident prediction that the bay offered hurricane forces a release valve. In fact, there was a sympathetic rising of the bay in response to the height of gulf.

So another record was being broken as night descended on Galveston and as the storm gained in intensity. Along with the deadly low pressure and the artillery-like winds it inspired, flooding of the island—from both directions, the gulf and bay—was reaching a height never before recorded.

As the Cohen family sang Gilbert and Sullivan in the parlor, water rose all the way over their Broadway porch while plaster crashed to the floor with each shock of wind. Doors at the school building where Annie and Ed McCullough and their families had fled kept blowing open, no matter the weight of all the big men trying to hold them shut. Uptown near the harbor, the streets and sidewalks lay below deep, swiftly flowing rivers. Structures as big and solid as the Levy Building were rocking and booming in the wind.

The passengers stranded upstairs at the railroad station had watched a dead child float below them in water. The collapsing second floor above Ritter's had killed five people.

And yet, as bad as things were, when Galvestonians retreated to public buildings and upper floors that evening, and when they lit their kerosene lamps against the darkness, they could be forgiven for hoping that what was banging their windows and shaking their foundations was hitting its peak—that the storm would soon pass. They had reason to hope that the gruesome extent of the damage, already well beyond anything they'd seen before, would be the most they were doomed to suffer.

That hope was in vain. The north wind that had been pushing back against the tide all day was letting go. The wind would stop holding the water back.

Soon it would turn all the way around. Roaring out of the south at record-force gales, it would get behind the biggest wall of water yet. No longer simply failing to hold the water back, the wind would now give the wave a huge assist. It would hurl that wall of water forward into town.

What happened then could not have been imagined by anyone in Galveston that night.

At the school, Annie and Ed McCullough and their families, trying to wait out the storm along with dozens of other distraught refugees, heard the people upstairs calling out for their families in the dark, weeping and praying. That's why they'd decided, at first, to sit in the hallway running north-south, not facing the storm's winds directly. They also hoped maybe the walls on that orientation would hold out longer.

But there was pandemonium in that hallway—hardly anywhere to sit. Men still strained every muscle to hold the doors closed against the surges of surf and the record-force winds. But the wind was too much for them: water kept flooding in, and it was getting deep.

Ed told Annie and the others it would be wiser to move to the east-west hall. The McCullough family began moving from the wet floor, but they were about to witness the most shocking event of their lives.

Clarence Howth, the lawyer who had begun the afternoon smoking a cigar while watching water rise from his window, was feeling a lot less calm now with evening falling. There was an eight-foot fence around his garden, and he couldn't see any of it—it was all underwater.

And as he'd gazed out his windows that afternoon, he'd watched his chicken coop falling, full of chickens, into the rising sea. He'd watched his chickens drown.

Meanwhile, slate tiles torn from his roof by the wind were flying wildly away. The resulting leaks brought in the driving rain. All of Marie's beloved upholstered furniture, lace curtains, and pillows were getting drenched.

As Howth peered nervously out his windows into the darkness of raging storm and evening dimness, he saw no houses to his east. They had just been there, lining the street as always. Now they were gone.

He looked westward. All the houses that way were gone too.

His house stood alone. His sick wife and new baby were on its third floor.

Shutters began falling from the house. Then windows started breaking, throwing glass around the inside. The wind was out of the east now, and the east side of the house would go first, Howth realized. His wife and baby were on that side.

He got busy. He went to the east room on the third floor where his wife and newborn baby were lying. Mrs. Howth had been blissfully unaware until moments earlier how bad things were getting. Her father, a doctor—a former Confederate medical officer—had been attending his daughter and the infant in her room, along with a baby nurse. Mrs. Howth's brother was staying in the house as well.

Clarence Howth called both his brother-in-law and the family's maid up to the third floor. All five carried Mrs. Howth and the baby, on a mattress, to a room on the west side of the house, away from the direct onslaught of wind.

But this side offered small comfort. They were all sure now

that the house was doomed. For those leaks in the ceilings were pouring rainwater now, and ocean waves were splashing against the windward windows, all the way up here on the third floor.

Howth paced from his wife's bed to the window and back in an agony of suspense. How long could these windows last? Might the storm abate before the house fell? Might this be the one house in the neighborhood that would survive?

Then they heard the worst crash of all. The windward side of the house had caved in. It fell, exposing them to the rain, sea, and wind. Quickly they retreated upward, again carrying the mother and baby.

They gained the drenched attic. They could go no higher.

In the upper story of Daisy Thorne's apartment in Lucas Terrace, the crowd of neighbors was now packed into the parlor. Evening was falling, and something was bumping the floorboards from underneath. Daisy realized with horror that it was the first-floor furniture, bobbing in the water down there.

Then came the shocking arrival of still more neighbors. Not from below, this time, and not from down the street—that could never happen now—but from upstairs, on the third floor of a nearby section of the same apartment house.

The most surprising fact about this arrival was that these neighbors got into the Thorne apartment at all. They'd been in the adjacent section of the building. It was an elegant feature of the Lucas Terrace Apartments that each section had its own entrance, with no doors offering communication among them. With the buildings' sections cut off from one another, how had these escapees possibly managed to arrive at the Thornes' door?

Necessity, luck, and brains. This group too had been seeking refuge on the second floor of a two-story apartment in that adjacent section; above them was a third-floor apartment. Mrs. McCauley, the woman who had been painting the romantic, russet-haired nymph with Daisy Thorne as her model, was there with a number of other women; the painter's husband, J. P. McCauley, was there too. And he was paraplegic.

They'd huddled there as a group, hearing a rumble, then a crashing splash. Part of the building's lower wall had failed. The sea had taken it down.

That meant the whole facade and wall on that side of the apartment building would soon go. They would be swept out with it into the surf. They had to leave.

At first, they ascended. They found their way up the public stairs. They checked the door of the third-story apartment above. It had been vacated and left unlocked. They went in.

Still, they remained on the side of the building that was collapsing. Being up here wouldn't save them. Somehow they had to get to the other section of the building.

In the apartment, they found an ironing board. They pushed open the door to the third-floor porch. Ducking the sheets of rain that swept across the sea below—that sea, which now hid their building's first floor—they used the ironing board as a plank. It spanned the wet, open air between this upper porch and the upper porch of the apartment above the Thornes', on the other side of the building.

In raging wind and rain, the women carried the paralyzed man across the ironing board. They all gained the other porch and entered the empty apartment there. They left it and quickly went down the stairs to the second story. There they knocked on the Thornes' second-floor door and were welcomed in.

But despite the astonishing resourcefulness of the group's self-rescue, the news they were bringing to the crowd sheltering on the Thornes' second floor was terrible. If one side of

Lucas Terrace was collapsing into the counterpunching ebb and flow of the surf, the rest of the building must soon follow. There was nowhere left to go.

The furniture below was still banging the floor. Soon the furniture would stop banging. It would be pressed, with gigantic force, harder and harder against the second-floor joists. The floor wouldn't stand it long.

Daisy had been planning big changes like a wedding, an end to her employment, and a move to Houston. Now, inside and out, her house was falling into a raging sea. She was sure she'd be going with it.

In the north-south hallway of the school, Ed and Annie McCullough and their families arose from the crowded floor, trying to get out of the rising waters where they'd been leaning against a wall, when lightning struck the building's chimney. Then came a long crash.

Annie froze. She turned to look back at where she'd been sitting.

The flue had collapsed without warning and brought down the entire hallway brick wall. Fifteen people were killed instantly, buried in the blink of an eye under a mountain of debris.

Annie stared in disbelief. Seconds before, she'd been sitting there too, right beside people who were now only buried corpses.

In the Howth attic, the family crouched together. Marie's brother knelt beside the new mother and prayed for her life. Gulf waves began smashing repeatedly against the attic window.

The old doctor was busy pushing an old cot against that window. He held it with all his strength. He was trying to support the window and keep the sea out, hoping he might just outlast the peak of the storm.

"Papa," his daughter called from her mattress. "Are we going to die?"

He reassured her: "No, daughter. It's almost over now."

But Howth felt strongly that they all had but a few minutes to live. He knelt by his wife and took her in his arms.

"Good-bye, darling," he said. "We will meet in heaven."

The doctor meanwhile must have seen his own doom coming. He shouted a final instruction. "Stay with the house as long as there's a piece of it!" he yelled. "If she stands this five minutes, it will all be over!"

The sea crashed through the attic window. The doctor disappeared. Howth was thrown by the blast away from his wife and baby and was underwater, plunged below the churning surface.

He knew it was over. He actively tried to make the end come quickly. Clarence Howth opened his mouth. He sucked in as much water as he could.

"The Lord is my shepherd; I shall not want . . . "

Daisy Thorne listened to her mother read aloud from the family Bible. The huddled crowd in the apartment listened over the roar and scream of the wind, seeking solace in the ancient words of the psalm. The floating furniture from downstairs continued to bob and bang on the floorboards.

Mrs. Thorne reached the end of the psalm: "Surely I shall dwell in the house of the Lord forever." The people began praying silently. Outside, waves began smashing against the second-story windows. The whole building was almost underwater.

Then a crash, startling everybody. The entire adjacent section of the building had fallen into the gulf.

In Mrs. Thorne's bedroom, Mr. McCauley, the paraplegic, was lying on the bed, his wife beside him. The crowd went into the bedroom and deliberated desperately. Daisy's bedroom, facing away from the gulf, seemed the safest place now. The crowd pressed that way.

But Mr. McCauley didn't want to move. His wife stayed beside him.

The crowd pushed into Daisy's room, filling it to the walls. At that moment, an adjacent public hall collapsed, sending bricks crashing through the room's windows, breaking the glass. Daisy ran to her mother's room to get pillows: she hoped to use them to protect her mother and aunt from flying glass and brick. She pleaded with the McCauleys to move from that room. They would not.

As Daisy ran back to the people in her own room, the parlor wall fell. She was looking straight out at the crashing water.

There wasn't much building left. Now big timbers crashed into the only wing that—other than the stack with Daisy's room—remained of the sturdy Lucas Terrace building. Through the broken panes in her room, Daisy looked across to see that entire wing of the house fall. The young woman had the fleeting impression that it exactly resembled a house of cards.

Then Mrs. Thorne's bedroom fell into the water. Both of the McCauleys were swept, together, straight into the churning waves.

Arnold Wolfram's family had begged him not to return to work after lunch. The storm had already been raging then, the streets filling with water. But Arnold, still stubborn, had felt he was needed at the store, so he'd gone all the way back uptown, to the store at Twenty-Third Street and the Strand, twenty blocks away from home—despite the fact that streetcars had stopped running, and the only way home would surely be on foot. The former ranch hand, now such a staid salesman and bookkeeper for a grocer, was fearless.

And now he doubted he would ever see his family again.

He'd finally realized, by early afternoon, that this rain was a true deluge, that the whole city must be in serious danger. He'd raced to close up the grocery store. He'd started into the streets, where the water was already rising quickly. He could see it filling the gutters and rising toward the sidewalk.

Arnold was carrying a package containing a new pair of shoes, bought that day. And now, as he tried to make the twenty blocks back to his family, he grasped for the first time just how serious the situation was.

It wasn't just rising water. As Arnold waded slowly, alone, against that rising current, he saw that he was in danger from above. Roof tiles had begun flying through the darkening air. Big pieces of glass windowpane were careening by: they turned on their jagged angles at high speed.

Arnold ducked. He was sure he'd be knocked unconscious or decapitated at any moment. He'd never seen anything like this wind-driven barrage.

He stepped up into the shelter of a doorway, and from the package he pulled his new shoes. He threw the packaging into the raging stream. Using the laces, he tied the shoes to his head.

Now at least he had a helmet. Turning into the wind again, he resumed his tortuous wading.

That's when he saw the child. Down in the rushing water a boy, unable to stand, had been swept along the streets from somewhere and was now, Arnold saw, spiraling strangely in place, around and around in the water. Unable to get up or swim away, the boy seemed stuck in a rapid circular motion.

Arnold realized what was going on. The kid was circling a deep storm drain. He was being sucked into the drain's vortex. Soon the kid would be pulled, helpless, all the way underwater.

As fast as he could, Arnold splashed awkwardly over to the boy. Just as he arrived, the boy seemed about to go under for good, but Arnold grabbed for his body. He got a grip. He yanked, breaking the vortex.

Arnold pulled the boy out of the drain and into the deep stream. There at least the boy could stand up.

And now Arnold recognized this kid: a Western Union messenger, about ten years old, who lived near the Wolframs. Then and there, Arnold resolved not only to get home himself but to get this boy safely into the Wolfram house as well. Over the howl of the wind, he shouted reassurance.

And he told the kid to take off his shoes. There in the deep, flowing streets, sheltering as well as possible against the rain and wind, the messenger boy did as instructed, and Arnold tied the shoes to the kid's head, just as he'd done with his own. Now they both had some protection from flying objects.

Together, they turned into the drenching wind. Arnold and the boy began their fight to get to the Wolfram home.

Sister Elizabeth Ryan finally made it back to St. Mary's Orphanage on the beach past the west end of town. The maintenance men that Mother Superior Camillus Tracy had sent to find the missing nun made it back too. Sister Elizabeth had been at St. Mary's Infirmary in town, collecting supplies for the orphanage. Having gathered groceries and other supplies in her wagon, she'd been ready to return, but Mother Gabriel, head of the infirmary, had been watching the weather and was concerned.

That was late in the morning, when some people were still treating the storm as a lark. Yet Mother Gabriel suspected it might be unsafe for Sister Elizabeth to travel alone by wagon all the way out to the westward beach. She suggested that Sister Elizabeth wait out the storm at the infirmary here in town.

If she did that, Sister Elizabeth objected, the children would get no dinner. She was needed at the orphanage.

She declined Mother Gabriel's offer. When the wind started shaking down bathhouses and telegraph poles, and the tide was starting to flood the streets, Sister Elizabeth was driving her wagon westward out of town.

And she did make it. While the orphanage was right on the beach, just past the high-tide line, and directly facing the gulf, its two buildings were brick and stone. It was a tall, imposing, two-part edifice that suggested nothing if not stability. A high wall of sand dunes seemed to protect the two-story structures from the gulf. Salt cedar trees grew in the dunes, screening the orphanage. The ten sisters had long been making those building at once a refuge and a home for the ninety-three children who lived there.

But by the time Sister Elizabeth returned, the orphanage was anything but a refuge.

The dunes had quickly been eroded by that tide rising higher, against the north wind, than anyone had believed possible. The salt cedar trees, shallowly rooted in the dunes, collapsed within minutes. Like every other building on or near the beach—just like Daisy Thorne's building, Lucas Terrace, on the opposite end of town—St. Mary's Orphanage was very quickly surrounded by crashing surf. By nightfall, the surf was crashing against the second-story windows.

Using lamps and candles, Mother Camillus, Sister Elizabeth, and the others had gathered all of the children on the second floor of the newer building, the girls' dormitory, farther from what had once been beach and was now raging gulf. Mother Camillus feared the worst. The older building would collapse first; then would come the building they were all in. Her only thought was to keep the children as calm and unafraid as long as possible, and keep trying to save them.

She led the children and the other sisters in song. They sang "Queen of the Waves." It's a folk hymn, originally sung by French sailors. The hymn asks the Virgin Mary for protection from storms at sea, and its lyrics—which the orphans and nuns of Galveston sang in English translation—described their position exactly:

> *See how the waters with tumultuous motion*
> *Rise up and foam without a pause or rest. . . .*
> *Help, then sweet Queen, in our exceeding danger,*
> *By thy seven griefs, in pity Lady save;*
> *Think of the Babe that slept within the manger*
> *And help us now, dear Lady of the Wave.*

As the children's voices rose against the booming surf and screaming wind, Mother Camillus's fears for the front building came to pass. They all heard the crash. The boys' dormitory had fallen.

Quickly she had the other sisters and the maintenance men tie the smaller children together with clothesline in groups of six or eight. Each group was tethered to one of the adults. The Mother Superior's hope was that in a deluge, the adults would be able to swim, and the children would not be lost.

And now, as the children kept singing, it happened. The

water and wind attacked the second building. It broke glass windows and undermined the floors. Then the roof fell.

The entire building toppled into the wild waters, just as the first building had. Mother Camillus, Sister Elizabeth, the maintenance men, and all of the orphans and other nuns, tied and untied, were washed into the sea.

Arnold Wolfram and his ten-year-old companion, the Western Union boy, were still fighting their way through waters rising so high as to become nearly impassable. Arnold was holding the kid's hand tightly.

And yet, shouting over the gale, he'd also made the youth understand that if one of them fell, they must let go, not bring one another down. The boy understood.

They had reached Broadway. That street, hours earlier a lovely esplanade on high ground, was invisible now, a raging sea.

They let go of each other. They swam down the street. Wires and poles were snapping, collapsing, flying. Yet somehow they made it across and grabbed the cast-iron fence of what had been a garden, outside the Galveston Artillery Club.

Standing on sidewalk again, with water now up to the boy's armpits, they worked their way slowly along the fence, holding fast to each rail.

At the corner, they came upon a man clinging to the fence. He was fading fast, exhausted; his grip on the fence was weakening.

Arnold and the boy moved as quickly as they could along the fence toward the man, thinking they could help. But before they got near enough, his grip failed. While they held to the fence and watched helplessly, the water swept him out of sight.

Arnold Wolfram was sure he and the boy were doomed too. But there was nothing to do but keep going.

Upstairs at Cassie Bristol's boardinghouse it was dark now. As little Louise and her family waited there, they felt the whole house shaking terribly, despite the holes Cassie had chopped in the ground floor.

The family wanted light. Through the rain-beaten windows, they could see dim lights in other upper rooms. People were using kerosene lanterns to bring some comfort to their places of terrified retreat.

But Cassie would not use a kerosene lamp. The awful shaking of the house might knock it over and set the place on fire.

Ever resourceful, instead she put a drum of lard, rescued from the kitchen, in the center of the room. She turned it into a lamp by plunging a stick into the lard, draping a piece of lard-soaked cloth over the stick, and lighting it.

The cloth burned slowly, consuming fat. Louise and her family, waiting in terror as their house swayed and water rose to the windows, at least had a mellow, stable light.

Then Louise's older sister Lois let out a scream.

What Lois was pointing out, to Louise and the rest of the family, was truly terrifying: the entire wall of their room was moving in and out with each blast of wind. The wall separated from the ceiling; it drifted outward far enough to give a full view of the rainy night sky. Thick clouds fled across the face of the moon.

Then the wall swung slowly back to join the ceiling and close off the view.

The room was falling apart. This called for evacuation, Cassie decided. The plan, as she delivered it quickly to her children, was this: she couldn't swim, but her sons were good swimmers. They could swim across while pulling a mattress; Louise, Lois, and Cassie would hold onto the mattress and use it as a float.

It was a desperate plan. Their destination would be a neighbor's house, still standing across the street—now ocean—with light showing on the second floor.

They got the mattress and readied themselves. When the wall left the ceiling again, offering them a way out, Cassie shouted, "Let's go now!"

But they all flinched when faced with the reality of leaving the house and embarking on the raging dark water. The sea was wild, and heavy objects shot randomly out of it. Overhead the sky was full of flying debris.

The wall closed again. They'd missed their chance. The house boomed and shook.

Then the wall opened, and Cassie shouted the same instruction. But Lois said "Wait!" Failing again to make their move, Louise and her family crouched, poised to flee but fearing to flee, in the room with a wall that seemed about to float away into the sea.

Joseph Cline had left John Blagden in charge at the weather station in the Levy Building. He told Blagden that things in the station, at least, were about to calm down. Certainly there would be no more phone calls.

And Joseph returned, still soaked, to the street. He again faced the gales and rivers that uptown Galveston had become.

At first his destination, as he splashed more than a mile through those rivers, fighting those gales, was the gulf beach near Isaac's house. He had it in mind to warn anyone he saw there that the peak of the storm was coming soon.

But he decided to take shelter at his brother's house. That building seemed capable of withstanding any assault—even these winds that, as Joseph waded his way downtown, kept knocking him off his course by many feet. It was true that the

Cline house stood very near what had once been the beach and was now a deep and raging gulf. And Joseph himself had seen water covering the lawn that very morning, before things had gotten so bad.

And yet as he neared the house, shouting at anyone he saw to seek the higher ground of Broadway, Joseph looked upon his brother's home as "a lighthouse built upon a rock." In all previous storms, it had easily weathered the worst. It was built for that. He had to get there.

Isaac Cline, for his part, had already retreated home. There was nothing left to do after giving Joseph the desperate message to Willis Moore in Washington. Galveston was cut off from the outside world. The storm was only increasing in violence as evening fell. The elements themselves, Isaac felt, had terminated his services as a weatherman.

So he began wading homeward to Cora and the three girls.

Timbers flew in the air around him. They drove themselves into house walls, splitting the paling and weatherboarding. Entire homes were being opened to the wind and rising tide.

But Isaac's house was new. It was built for stability in storms. It had withstood all previous weather. Like Joseph, Isaac believed it was the place to be.

Isaac and Joseph Cline were hardly the only people in Galveston who believed in the security and stability of the Cline home. When Isaac finally arrived there, he didn't find only Cora, ill and weak with her pregnancy, and the three Cline girls. He also discovered fifty neighbors.

They'd come there for safety. Among them were the builder

of the house, along with his family. He knew what kind of structure he'd put up.

And yet if Cora hadn't been so weak and frail with this fourth pregnancy, Isaac would have wanted to move her and the girls up off the beach, toward Broadway, maybe even all the way to the Levy Building itself. However strong Isaac and his builder knew this house to be, higher ground seemed even safer. But getting Cora out into the storm seemed riskier than staying put.

By the time Joseph arrived, the water outside their home was waist deep. All of the houses directly on the beach were down. Inside, Joseph found the place milling with increasingly distraught families. And he found his brother Isaac at Cora's bedside.

Hoping to avoid further terrifying their family and the fifty refugees, the Cline brothers stepped together onto the porch to discuss their limited options. The water was nearly all the way to the top of the porch steps.

And as they stood there, weighing the risks of staying in the house or trying to escape uptown, the elements answered the question for them. What Isaac had expected came true.

The north wind was no longer a north wind. It had been shifting. And now it made its decisive move.

It came out of the east. That is: the wind no longer held back the rising gulf.

Standing on the porch, the Cline brothers watched the water rise four feet. It took four seconds.

So there was no chance of leaving. The porch was underwater. The Cline brothers turned and retreated inside.

There they faced Cora, the three girls, and a house full of refugees. Isaac and Joseph were the experts. Everybody wanted to know what to do.

But the water, having risen four feet in four seconds, followed the brothers inside. Trying to restrain the growing panic, the brothers ushered everybody to the second floor.

There, they had to choose a room to pack everybody into. Somewhat counterintuitively, they picked not a room away from the wind, but a room on the windward side of the house.

That wasn't an accident. Joseph and Isaac had quickly come up with a theory—even a kind of grim plan. Should the blasts of punishing wind finally push the house over, the Clines reasoned, they would want to be on top of, not under, what would then be a house flipped on its side and floating. They needed to stay on the wind side, which would become the top side after a capsize.

The house itself, or parts of it, might then serve them as a raft. And now the likelihood of even this strong house going over was starting to seem high. From the packed second-story room, shaken horribly by the direct wind, the terrified crowd could see other houses that had come down and split into pieces. Those houses churned, becoming massive debris in the swollen tide and heavy rain. The Cline home, the brothers felt, was soon to join them.

Arnold Wolfram and the messenger boy had reached Broadway and Twenty-Fifth, site of the monument to the Texas heroes of Fort San Jacinto. Parts of buildings and other giant debris, brought uptown by the gulf water, had started piling up against the statues there. It was getting dark now.

The duo pressed on. After many blocks of navigating the rising seas, they were actually nearing Wolfram's house.

A few of the houses along Broadway seemed to be withstanding the storm so far. Arnold recognized the home of a friend and, looking up, could see through the windswept rain that the man was inside the house. Arnold and the boy waved and yelled for entry, maybe through a window.

In the dark and the wind, the friend didn't see or hear them. There was no time to keep calling out; there was nothing to do but move on.

Fording yet another river of a street, Arnold and the boy neared another friend's house. In shoulder-high water now, Arnold hoped to try again here.

Yet even as they approached Arnold's friend's porch, a mighty current suddenly grabbed them both and swept them off their feet. They went all the way underwater.

Arnold, pushed with immense force, banged into a tree. Still underwater, he did his best to grab it. He got a grip and held on desperately against the current.

Using the tree, he pulled himself upward to get out of the water. He clung to the tree as the water rushed by and looked about for the boy.

And miraculously, the kid was clinging to the tree too. He too had hit it, had pulled himself out of the water, and was hanging onto a branch.

This tree wasn't big. But it had thick branches that would bear human weight. With their last strength, Arnold and the boy hoisted themselves out of the water. They climbed into the fork of those branches.

There, just above the rushing tide, they huddled together, gasping for breath and sheltering as best they could from flying wood, slate, and glass. For now, they'd gone as far as they could go.

Packed into the rocking windward side of the Clines' second floor, Isaac, Joseph, Cora, the girls, and the neighbors were in the dark now. As they waited, looking out the windows, they saw that they were really alone. Almost every other house in the neighborhood had gone into the water.

The Clines' own porches, upper and lower, front and rear, had been torn away as well. The water was now at least fifteen feet high.

Joseph Cline was certain the house was doomed. He urged the assembled crowd to be ready for collapse. When it happened, he told them, try to get out the windows, climb on top of whatever house was left, and then ride the house like a raft through the storm.

He was in an altered state of strange calm. He felt that he would live. He and Isaac had an uncle who had survived a shipwreck by drifting on a plank, and Joseph believed they might do the same.

Others in the crowded room were anything but calm. Some were kneeling in prayer. Some were running around pointlessly, in total panic. Some were crying, some wailing aloud. Most, strangely enough, were singing.

Isaac was meanwhile watching the progress of debris in the wild water outside. Other houses were down because the wreckage of buildings had itself become a cause of further destruction. The waves tossed ruined timbers and roof structures again and again against standing walls, driving down walls and roofs together. Then the newly fallen wreckage added greater and greater power to an assault on the standing houses.

Finally there seemed to be nothing but wreckage. It seemed to Isaac to be hanging around his house—the last house standing—as if to menace it. In a newly horrible development, the relationship between the gigantic debris and the raging sea was changing.

The wreckage was forming a high dam. It held the water back so that its level was rising monumentally higher even than before. Thanks to this dam effect, caused by the destruction it had already wrought, the sea at the Cline house was soon towering twenty feet above the ground—ten feet higher than the storm tide itself.

Then, as Isaac watched, an iron train track appeared in the water. Here was a monster. He saw the trestles emerge from the sea, held together by track, and come shooting, an ideal battering ram, toward his house. As it came, the track revealed its length. It was a quarter-mile long.

This was the same trolley track that, on happy days, had run along the beachfront over the gulf water. Underwater for hours, the track had finally been torn loose from its moorings by the surging force of the water.

The sea had put the track on a rampage. It was speeding straight for Isaac's house. The swells brought the track through the high water. Squarely, at storm-tide speed, it smashed the wall of the Cline house.

It pulled back with the swells. It came again. It smashed the house. It pulled back. It came again. Again.

As gigantic iron battered his house, Isaac went to Cora's side in the center of the room; with them huddled their youngest girl, the six-year-old. The entire house made a loud creaking noise. Then it started to fall, spilling the people about the room.

Joseph was standing by the window when the whole house started moving. With him were Isaac's other two daughters.

Joseph executed his plan. He grabbed the hand of each of the girls, turned his back to the window, and lunging from his heels, used his back and shoulders to smash window glass and break through the wooden storm shutters. As the building slowly turned, capsizing, Joseph and the girls together went through the open window, toward the water.

Galveston's deluxe beach hotel, a big tourist draw in the 1890s and a landmark of America's emerging Gilded Age metropolis on the Gulf of Mexico.

One of Galveston's sturdiest buildings, the Gresham Mansion withstood the storm and later became the bishop's residence.

Modern Galveston before the storm: An electric streetcar comes up Market Street.

Bird's-eye view toward the gulf, 1894.

"Not a hurricane":
Willis Moore, director of the
U.S. Weather Bureau. NOAA

Portrait of the meteorologist as a young man:
Isaac Cline. NATIONAL ARCHIVES

One of the oldest weather
instruments, measuring wind
speed. NOAA

A basic hygrometer:
Relative humidity is
measured by "dew point."

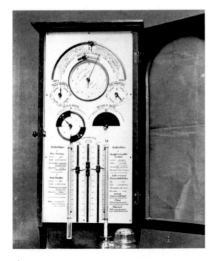

State of the art: A late-nineteenth-century
U.S. Signal Corps barometer, made in
England. LoC

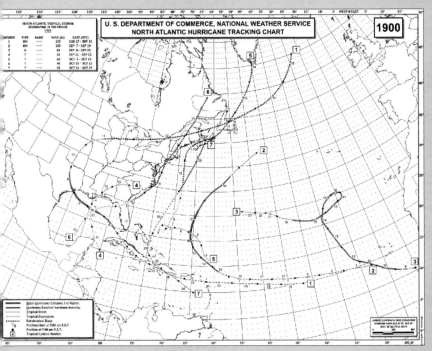

Storm tracking: For the actual path of the storm of the century, see line 1. NOAA

A hurricane "eyewall." NOAA

Eye and eyewall. NOAA

"Galveston's Awful Calamity":
Large chromolithographs dramatized disaster for an eager public. LoC

After the storm. LoC

Tracks on Avenue A,
near Twelfth Street. LoC

"The wreckage
had created a new
landscape. . . . There
seemed to be no city
left at all." LoC

"The ship seemed to have been
tossed out of the bay." LoC

On the pile. LoC

Seeking bodies. LoC

"Lying about the vast desert of wreckage was a multitude of corpses." LoC

The white city on the beach: U.S. Army tents as homeless shelters. LoC

Joseph Pulitzer's evening "yellow press" paper, fighting its headline war with William Randolph Hearst's *New York Journal*. LoC

Posing on the wreckage—a symbol of hope and resilience.
LoC

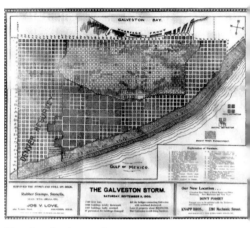

Charting the devastation.
GALVESTON HISTORY CENTER

Seawall under construction.
LoC

Galveston resurgent: The beach draws swimmers, the seawall draws crowds.
LoC

". . . slurry pumped from the dredgers in the canal came blasting out of those pipes. It flowed into empty spaces under the buildings, filling the space and leveling itself." GALVESTON HISTORY CENTER

Isaac, Cora, and their six-year-old were meanwhile tossed like the lightest of objects by the impact of the trestle, the turning of the house, and the entry of the sea. They were sliding fast into a wide chimney. Down the chimney they tumbled, all the way to the bottom of the capsizing house, then out of the ruined house and farther down, to the very bottom of the violent sea twenty feet deep.

They were underwater.

Trying to rest in the fork of the tree branches above the water, Arnold Wolfram began to see that his and the boy's situation was becoming precarious. It was totally dark now, but despite the rain and cloud, the moon was faintly visible. In its dim light, they could see not only the sky, but now also the dark, churning sea just below them, full of gigantic objects. The debris swirled, plunged, and launched from the fast water.

The wind and the ocean had already brought down hundreds of buildings toward the gulf side. The water was now pushing all that debris chaotically uptown toward the higher ground. Arnold and the boy could see, thanks to the dim moonlight, the deadly battery that the tide was sending at them—they could at least dodge and duck as they waited in the tree.

But the tree had now begun stopping the wreckage in the water. Debris pounded the tree and then piled up against it with immense weight. Arnold thought the tree might be battered to pieces or pushed over.

And he and the kid were freezing cold now, soaked, weak from hunger and sheer exhaustion. How long could they possibly hang on?

As Arnold considered what to do, he and the boy saw something out of a nightmare. A house roof came bearing straight down on them, through the water, and trying to ride it like a raft were a man and a woman. The roof came crashing into the tree where Arnold and the boy perched.

The roof split in half. The man was carried away into the darkness.

The woman's section of the roof got stuck, bobbing against other wreckage near the tree. Screaming for help, the woman reached out for Arnold and the boy.

They tried to get her. The kid held tight to a branch; Arnold took his other hand and leaned out toward the woman, extending his hand toward hers. Arnold and the woman stared each other in the face as she reached toward him as far as she could.

Her chunk of roof broke free of the wreckage. The woman's face grew anguished, and she shrieked as the roof jumped away from the tree and plunged. Arnold could only stare as the woman went under and disappeared.

At nearly the same moment, a long rafter beam came bearing down on the tree. It too stuck to the branches.

After ducking its impact, Arnold picked up his head and saw something amazing.

The beam's other end had landed on the upper porch of his friend's house—where they'd been hoping to seek shelter before the current had sent them into this tree. That house had remained standing. Unlike others, it still looked fairly solid.

The rafter seemed to be tightly stuck at both ends . . . one side in their tree, the other on the upper porch. There was a slim chance that it would hold long enough to serve as a bridge from the tree to the house.

By now, Arnold and the boy were thinking in unison. Their tree was swaying, about to break under the weight of debris. Yet even as Arnold moved to lower the boy onto the rafter, the youngster had anticipated him.

As one, they were splattering their way, as quickly as they could, across the wet board toward the porch.

Joseph Cline was climbing with haste from the rail trestle to the topside of the overturned house. He still grasped Isaac's two little girls' hands and pulled them upward. They all gained the top and crouched there above the water.

The three were anything but safe. For now, though, they were alive. They crouched on the house, which had not broken up but was bobbing violently on the surf in wind-driven torrents of slashing rain. The girls clung to their uncle for protection from flying debris as the house rose and fell. The surface would climb a few feet above the rushing tide, point upward toward the dark sky, then plunge almost into the black water.

Riding this gigantic, bucking raft, Joseph was still thinking about all the people who must still be inside the over-turned house below them, full of water. There was nobody else on the swirling wreckage, nobody visible in the dark water. The window Joseph had broken, on what was now the top side of this floating structure, would be the only exit from that room.

So Joseph gently let go of the girls and put his head and shoulders down through the window. As the house rose and fell, he yelled "Come here!"—hoping to help any conscious victims locate the one exit.

He recalled that drowning people will grab at anything. He reversed position and dangled his legs through the window. He hoped to feel someone grasp them. Nothing.

With anguish, Joseph concluded that they must have all been trapped and drowned. That meant Isaac, Cora, and their six-year-old daughter too. He returned to protect his nieces on their rocking craft in the howling wind.

The danger had just increased: the house they were riding on was starting to break up.

Isaac Cline was gasping for breath, swimming in the water, pinned between two huge house timbers that had kept him crushed underwater so long that he had become certain of death. Somehow he and the timbers had risen. He was alive, breathing, fighting the currents.

Despite the driving rain and flying objects, there was a dim moon. Isaac scanned the churning waters now in desperate hope of seeing Cora, his daughters, his brother . . .

A flash of lightning gave a flash of hope. There was his six-year-old, floating on some wreckage. She was alive.

Isaac left the timbers and swam to her. He clung to his child and to the wreckage she floated on.

A moment later, another lightning flash: miracle of miracles. Isaac saw Joseph and the other two girls. They were riding on a piece of what had been Isaac's house. He grasped his youngest and swam with her to his brother and children.

By the time Joseph saw Isaac and the youngest girl in the water, the house that Joseph and the girls were riding had splintered into big pieces, and they were working hard to stay on top of what they had left. Joseph's heart leaped at the sight of his brother and the girl. He helped them aboard the precarious craft.

Quickly, Isaac told Joseph what had happened. Far under the water, Cora's clothes had become entangled with the wreckage. Isaac himself had lost consciousness when pinned. Somehow he'd survived. Cora, it was now plain, had not.

And yet that night, as they struggled to remain atop the debris that might yet save them, the full weight of their loss hadn't settled on the surviving Cline family. The children, Isaac noticed, weren't panicking or crying. They showed no signs of fear at all.

And the Cline brothers themselves, facing what they knew might still be their own deaths, felt nothing as recognizable as fear. This was all new. Their only idea was survival.

Survival. Galveston was sea now: there was no island. The bay had met the gulf, and as the Cline family spun, pitched, and coasted on their collapsing raft, all they saw to suggest a city were the tops of a few remaining buildings, here and there visible in lightning flashes above crashing ocean waves that spread on all sides as far as they could see.

And in the distances, they sometimes saw lights: signs of people in the upper floors of houses that remained standing, in the neighborhoods farther from what had been the beach.

So their battle began.

The howling sky was still alive with flying missiles that showered on their heads. Isaac and Joseph sat on the raft with their backs to the wind and the children seated in front of them, between their legs. The raft bobbed, circled, and surfed. Behind them, the brothers held some scavenged planks, hoping the wood might deflect any impact of flying objects.

And even as they fought deadly hits from the sky, this weak family raft, coasting the high waves, was slammed repeatedly by debris surging up from the water. Again and again missiles shot from the wind and sea knocked the brothers from their perch all the way into the surf. Each time they struggled back to the children, hauled themselves onto the raft, and resumed their positions.

They could never stay securely on one piece of debris. Their makeshift perches kept collapsing into the water, and they all climbed and crawled together to a more solid chunk of wreckage, which in turn became only a temporary refuge. At one point Isaac was slammed in the back by a flying timber and fell forward onto his face on the float. Yet somehow he got up again. Joseph meanwhile tasted salty blood. He put his hand on his face and then on his head. A long gash was running through his scalp: he'd been struck by broken window glass. And yet he felt no pain at all.

The Cline family could hear, above the roar of the wind and the surf, human screaming. The suffering of dying and injured people, mostly invisible in the darkness, was at times intensely audible.

Now a lightning flash abruptly revealed a small girl of about four, floating alone on some wreckage. They pulled her onto their own raft and placed her with the children.

Suddenly, Joseph feel a shock of terror. Even as the Clines were struggling to stay afloat, he saw a monstrous hulk, nearly a whole house, careening toward their modest raft. This huge hazard carried big piles of smaller debris before it. In seconds it would stove the Clines' raft in, batter them, and dump them all in the surf.

Joseph and Isaac leaped to their feet. Just as the hulk, bearing down on them at full speed, arrived at the raft, they grabbed its upper edge; they hung on it. Their weight brought the whole thing close to the water line, slowing it down and holding it off.

Keeping this monster in a strong embrace, the two men pulled its top edge as low as they could. They told the children to climb aboard the hulk. Joseph and Isaac scrambled up with the girls.

They all climbed to its top. That gigantic, deadly object now became, for a time, their new haven.

Then came something truly strange. Joseph's retriever—his favorite hunting dog, the family pet—suddenly came scrabbling out of the swirling water, climbed onto the raft, and shook himself dry.

Could this be random chance? Had the dog somehow searched for them in all the dark turbulence?

The retriever ran about the raft, sniffing each person in deliberate fashion. Then he ran to the edge of the raft and looked about urgently. Joseph had the unmistakable impression that his dog knew one person was missing.

Cora. Joseph thought his dog felt called upon to find her.

As the dog prepared to leap back into the sea, evidently in search of Cora, Joseph shouted to him to stay. The dog ignored the command. Joseph lunged to stop the animal.

No use. The dog dodged him, jumped, splashed, and disappeared into the waves.

Through all of those hours of weird and relentless struggle, the Cline family journeyed an incredible distance. There was no steering. There was nowhere to navigate to, no direction, no orientation. Trying to hold on and avoid further injury, they spun and coasted wherever the wind, the tide, and the waves sent them.

At one point, they realized they could no longer see any lights at all. That meant they'd been swept far out into the Gulf of Mexico. Much of the wreckage that had once been Galveston had come with them. They might, they knew, be many miles from what had been their city. They might be swept all the way out to sea; they might never get back. The storm raged on as they tried to hold their perch.

PART III

THE WHITE CITY ON THE BEACH

CHAPTER 11

TELEGRAPH SILENCE

WINIFRED BLACK WAS DISGUISED AS A BOY. THIS RED-HAIRED reporter for the Hearst newspaper chain and newswire service was thirty-seven, but she was still slightly built, and still undaunted. Winifred had always done whatever it took to get a story. This was starting to look like one of the biggest.

She had to get into Galveston. No other reporters had been in yet. She had to be the first.

She had her long red hair tucked up under a workman's cap. She wore men's shoes and a linen duster and carried a heavy pick, which she was trying to keep resting securely on her shoulder.

Hiding among a yelling, rowdy gang of men—they'd been recruited in Houston for the relief of Galveston—she hoped to board a boat and cross Galveston Bay. Night had fallen. As the crew shuffled toward the gangway, Winifred stayed between two huge men. She hoped she wouldn't be noticed. Her heart was pounding with fear and excitement. Police and soldiers were patrolling the mainland gangways. Guards stood at the bottom of this very one, looking people over as they boarded the boat.

The bay was off limits. Nobody could travel to the island without an official purpose. News reporting was most certainly not an official purpose. But as a reporter, Winifred Black had often drawn upon her talent for dress-up, make-believe, and dramatic gesture. Her trip to Galveston would be no exception, and yet on this assignment, she would find herself doing far more than reporting.

Clara Barton was meanwhile leaving Washington, D.C., on a train, riding in what was called a "palace-car." Miss Barton, as she was always known, was seventy-eight now; the venerable founder of the Red Cross was traveling the way the great industrialists and magnates did. A far cry from the spartan, dirty long-distance sleeper cars that ordinary passengers had to endure, palace cars were ornate, plush, and clean, fitted out with elegant comforts and staffed with servants.

And that's how Joseph Pulitzer—William Randolph Hearst's rival in the newspaper game—wanted Miss Barton traveling to Galveston. She would bring relief, as paid for by his own paper, the *New York World*.

If Hearst had Winifred Black, Pulitzer had Clara Barton. Among the entourage were not only about a dozen members of the Red Cross but also Robert Adamson, one of Pulitzer's top reporters. The train hauled cargo, too: medical supplies, mass quantities of disinfectant, and money to buy food, water, and further supplies.

Clara Barton had been witness to—really, up to her neck in—the most tragic scenes of destruction of the second half of the nineteenth century. The Battle of Antietam in the American Civil War, the Johnstown Flood in Pennsylvania, the Armenian massacres in Turkey: she'd built her organization by maintaining a fearless commitment to facing disaster head-on. She was tireless. She and her people worked until relief was achieved.

Now Miss Barton was rolling toward a scene in Galveston that she would later call more horrifying than anything she'd seen anywhere else, ever before.

By the time both Winifred Black and Clara Barton were making their separate ways—in their distinctly separate styles, each at the behest of a rival newspaper publisher—toward Galveston Island, much had happened both on the island and off to challenge people's imaginations. The horrors of the night of Saturday, September 8, 1900, were already matched by an ongoing nightmare on the island. It was beyond anything that could have been predicted. Both Winifred Black and Clara Barton were about to play important roles in an effort at relief and recovery that looked, at first, overwhelmingly impossible.

Sunday, the ninth of September, began, for those outside Galveston, with a terrifying mystery. In Washington, D.C., Weather Bureau Director Willis Moore spent that morning anxiously awaiting any news at all from his Galveston station.

He'd received reports from elsewhere on the Gulf Coast.

The storm was a hurricane. Moore knew that now.

But he hadn't heard from Galveston, and now he feared the worst. The last cable he'd received from the bureau office there—the one sent midafternoon on Saturday, against such great odds, by Joseph Cline—had told Moore that half the city was underwater. Dictated over the phone to the Houston Western Union office, that cable had predicted great loss of life; it had asked for immediate relief. And then, moments after the Houston operator wrote the message down, all phone and telegraph communication between Houston and Galveston had been lost.

Since then: nothing.

Moore had been able, however, to make contact with the bureau office in Houston. That city had been rocked hard by the edges of the hurricane, but it still had power and communications; it was standing. Houston's weathermen reported to Moore that they, too, could not establish contact with the island city.

So now Moore sent a telegraph directly to the manager of the Western Union office in Houston. It read simply, "Do you hear anything about Galveston?"

The answer was no. All telegraph lines south of Houston were out, but even more dismaying, the telegraph office in Tampico, Mexico—well to the southwest of the island and therefore not involved in the storm—reported that it too had no communication with Galveston.

That meant the problem wasn't between Galveston and Houston because of downed cable lines on the Gulf Coast. Galveston was cut off from everywhere else too.

On the official U.S. morning weather map, on the morning of September 9, 1900, the bureau had no choice but to mark Galveston, Texas, "missing."

With Galveston blacked out, the weathermen watched the hurricane move on. That's what hurricanes do.

Having made landfall on Galveston Island at the height of its fury, the storm roared on its northwest path straight for the west side of Houston. For miles the mingled gulf and bay tides flooded the flat, wide, long coastal plain of Texas.

In small towns, houses and phone and telegraph wires succumbed quickly to the wind and sea. Most of the people there, however, managed to make narrow escapes.

Houston itself, far enough inland to avoid the awful tidal flooding, got battered by wind and rain all night on September 8 and into September 9. But damage there was largely limited to property: a big factory and a big oil refinery became wrecked hulks; almost every church steeple suffered. Many Houston businesses lost at least their roofs.

Houston remained in communication with Washington. And despite the impossible weather and the property damage, Houston officials kept trying to reach Galveston by telegraph. But it was no good.

And still the storm moved on. In the ensuing days, this systematically cycling system of elemental violence, which had begun off the savannahs of West Africa, seemingly so far from North America, would make itself felt throughout much of the United States.

To Americans, there was a new sense of sheer bigness to their recently unified, muscle-flexing nation. The United States filled much of the continental map now, from coast to coast, from the Gulf of Mexico to the Great Lakes. By human standards, the country was gigantic—but those distances meant nothing to the hurricane of 1900.

Starkly different American climates made no difference, from seacoast to farmland to bayou to mountain prairie. Nor did the people's different accents, from Down East Maine to the Chicago streets to the deep South. Nor the range of architectural styles, from big office buildings to Frank Lloyd

Wright's radically modern homes to bungalows to tarpaper shacks to farmhouses.

None of the classic American differences, the long American distances, or the political bonds that connected disparate people and places meant a thing to the hurricane. It dwarfed huge regions. What to Americans seemed a vast and sprawling land—this nation so recently sure of its size, power, and strength—served the hurricane of 1900 only as a playpen. Real size, real power, and real strength belonged, as always, to nature.

So when the hurricane traveled into Texas, it was slowed only slightly by its confrontation with the island of Galveston and the plain south of Houston. It kept going. Soon it drenched Oklahoma. Keeping up its cycling motion, with winds diminished only to about half of what they'd been at their height in Galveston—still tropical force—that hurricane brought driving rains to the upper Midwest as far north as Wisconsin. Lake Michigan, like the gulf before it, reacted to those winds with high, damaging waves.

Turning eastward at last, the storm swept all the way into New York City, raising hell along its path. In Manhattan, people couldn't walk the avenues in the winds. Signs blew off tall buildings. One New Yorker was killed by a flying pole. Waves crashed higher against the Battery than anyone could remember. Ships trying to enter New York Harbor lost course. Over in Brooklyn, trees were torn up and thrown down. And at Bath Beach, not far from Coney Island—a famous seaside pleasure park, at another corner of the country from Galveston's—a bathing pavilion collapsed into the raging sea.

So by the time this hurricane did finally move back out to sea—off Nova Scotia, having swamped New England fishing fleets along the way—many Americans had received at least a small taste of the enormity that had arrived in Galveston on that terrifying day and throughout that long, horrible night of September 8.

Yet nobody, in Texas or anywhere throughout the country, could have imagined the effect that force had on Galveston. Until they saw it, nobody could believe it.

Well up the Mississippi from the Gulf Coast, the Western Union headquarters in St. Louis was overwhelmed on September 9 with panicked inquiries about Galveston. Western Union quickly organized teams of linemen and operators and sent them by train into stormy Houston. The hope was to get the workers from Houston to Galveston to begin restoring the island's communications and get information.

People in Houston with relatives in Galveston were meanwhile beginning to panic. Rumors flew. The city might be half destroyed. There might even be hundreds of dead. To sober minds, these rumors seemed wildly and irresponsibly exaggerated.

Still, Houston prepared quickly to go to help Galveston. With communication impossible by phone and cable, the only thing to do that Sunday was to travel to Galveston by train and see what was going on. A group immediately formed with hopes of getting into Galveston. Despite the damage the storm had brought to their own city, the citizens and officials of Houston were determined to send help southward.

That very afternoon, a relief train left Houston on the Santa Fe Railroad tracks, loaded with supplies and men. The train's immediate destination was Virginia Point, the site of the entrance to the rail bridge that spanned the bay to the island.

As the train chugged down the track, the day cleared up. It became bright and sunny, with a pleasantly gentle breeze.

But soon the relief train slowed to a halt. The tracks were gone. This was still well inland, more than six miles from the point where the bridge began. The team couldn't yet see anything of Galveston, of course. What they could view, here on the coastal prairie, was shocking enough.

The plain was strewn with dead bodies.

Along with those corpses, huge pieces of lumber were littered about the flat ground as far as the eye could see. Roofs. Packing trunks. Pianos.

From the stopped train, the relief team saw a whole steamship. It was wrecked—all the way up here on land. They stared, amazed. The ship seemed to have been tossed out of the bay.

As they observed this bizarre and horrifying sight, the relief team on the halted train was further startled. Two men were hailing them.

Exhausted and bedraggled, the men climbed aboard— refugees from Galveston. They'd been swept off the island, they told the Houston team, and had been swept all the way across the bay.

The team welcomed them onto the train, and as this first, failed effort to reach the island city began riding back up the track toward Houston, the men from Galveston began telling a shocking story.

The relief team listened, astounded, to this first report on the night of horror. The survivors estimated that there might be as many as 500 dead in the city. The Houstonians were sympathetic, but they took that to be exaggeration— understandable enough, given what these poor men had been through.

Back in Houston, the men were taken to the Western Union

station to begin the process of reporting the disaster to the nation. Another exhausted survivor reached Houston, also on the ninth. He reported an even higher likely death toll, maybe a few thousand.

That seemed even more unlikely. Still, people in Houston were beginning to get the idea that something truly awful had happened in Galveston.

In the Houston Western Union office, the manager, G. L. Vaughan, got busy. After taking reports from the Galveston survivors, Vaughan cabled the news to Governor Joseph Sayres in Austin. Next he cabled President McKinley in Washington.

Those two executives responded quickly. Governor Sayres issued a dispatch citing the most horrifying number yet: maybe 3,000 lives, Sayres announced, had been lost in Galveston.

President McKinley, in turn, cabled his sympathy to Sayres and offered immediate federal support. "Have directed the Secretary of War to supply rations and tents upon your request," the president told the governor.

Vaughan, the Western Union manager, meanwhile cabled an anxious Director Willis Moore at the Weather Bureau in Washington. He wanted to get Moore the latest information as quickly as possible.

"Loss of life and property," Vaughan said, "undoubtedly most appalling." Now Moore had at least some idea of what had happened to Galveston.

That message to Moore was sent late at night on September 9. So far, nobody had been able to get to the island. Nobody who hadn't been through that storm had seen what it had done.

Also on September 9, a group of men arrived in Houston from Dallas. These were representatives of the Phoenix Assurance Company, which held policies on much of Galveston's properties. Having read in the national morning weather report that the island was listed as "missing," the company wanted to know the extent of its losses there.

These adjusters were led by their boss, Thomas Monagan. He was persistent, resourceful, and well-organized, and he had company funds to spread around. Monagan was intent on getting into Galveston.

His idea was take a train, but on different tracks this time, in a different direction, as far as it would go, maybe all the way to the junction on the coast at Texas City. There, relief teams and insurance adjusters might board ships across the bay. They would reach Galveston by water.

But even while Monagan was organizing this next effort to get a relief train as close to the island as possible, U.S. Army General Chambers McKibben arrived in Houston with a squadron of soldiers. Lately of the Spanish-American War, McKibben had served as military governor of Santiago, Cuba. At once an able commander and a decisive administrator, he'd come to Houston under orders from the secretary of war in aid of Galveston.

The military man McKibben joined the civilian Monagan in getting the relief train outfitted—and as an officer, McKibben took over. An astute student of human nature, the general feared there would be thrill seekers and sightseers hoping to get into what he'd already concluded must be a horrible disaster area.

He didn't want that. He didn't like the idea of any news reporters either.

So the general gave the civilian Monagan an assignment. The insurance man was to ensure that every single person riding this train had an official U.S. Army pass, issued by McKibben himself.

Before dawn on September 11, with Monagan checking passes, the new relief train filled up with U.S. Army soldiers under General McKibben; Texas Rangers recently arrived under the Adjutant General of Texas General Thomas Scurry; volunteer relief workers; Monagan's insurance men; Galvestonians who had left town before the storm hit; and friends and families of those in Galveston, desperate to find loved ones.

Monagan watched those citizens closely. He felt for the worried civilians. A mood of tense anxiety pervaded the cars.

By dawn, the train was already moving toward Texas City on the coast. The day was beautifully clear, dry, and cool—one of those days when weather seems not only benign but actually hospitable, friendly to human efforts.

And yet as this train too neared the coast, the disastrous extent of what weather can do became shockingly visible to the passengers. Again the rail track ended, this time well before Texas City.

Worse yet: here on the mainland the standing water was at least two feet deep. The train came to a stop.

General McKibben barked orders. The soldiers, insurance men, and ordinary citizens removed their shoes, rolled up their cuffs, and detrained into the water. They started wading.

But as they slopped along toward Texas City, their journey became ever more bizarre and frightening. Corpses came floating by. The travelers could only hope that these deaths were among the few, but that wish dimmed as they kept splashing forward across the plain.

They saw not fewer corpses in the water, but more. Now there were household items. There were parts of houses. Dead bodies of domestic animals.

By the time Monagan arrived on the docks in Texas City, he was in a deepening state of shock.

Not so General McKibben: without even consulting Monagan, the general abruptly commandeered a small steamship and marched his soldiers onto it. Before the insurance man knew what was happening, the steamer cleared the dock and entered the bay. The general left the civilians behind.

Monagan began to realize he would have to make some unusual moves here on the Texas City docks if he was to get into Galveston. Stunned by what he'd already seen, Monagan was keenly aware that, as a mere insurance executive, his only real authority for entering the disaster area came from General McKibben. And the general had just abandoned him.

Still, Monagan was determined. He felt personally responsible not only to his company but also to these many civilians filled with terrible dread for family and friends on the island. One way or another, he needed to get these poor people across the bay.

He sighted a good-sized sailboat heading in toward the dock, a schooner. When it had tied up at the pier, Thomas Monagan—acting on no particular authority—announced that he was commandeering this boat on behalf of the Dallas and Houston relief efforts and the accompanying civilians, all of whom held U.S. Army–issued passes.

He shepherded nearly one hundred people onto the boat. Late that afternoon, its captain steered the schooner out into Galveston Bay.

Monagan had already seen dead bodies littered along the plain. And yet it was only on the bay that he realized the truly ghastly extent of the destruction.

General McKibben had commandeered a steamer. But Monagan's craft was a sailboat, and the wind, coming out of the south, was very light. The captain therefore had to tack, slowly and repeatedly, up and down the bay. The boat made the slowest, most tortuous kind of progress toward the island.

The trip took all the rest of that day, and it was during those long hours that Monagan and the others came to realize, with mounting dread, the scope of what they were really encountering here. The entire bay was clogged. There were broken houses, broken furniture, and broken personal effects everywhere in the water.

But that was nothing. There were also the bloated corpses of men, women, children, and babies. In every direction, Monagan and the other passengers on the boat saw hundreds of bodies in the bay. And so this agonizing ride, tacking up and down across Galveston Bay, became a slow-motion tour of heartwrenching, overwhelming loss.

With evening falling, the passengers at last arrived in sickened, terrified silence at the ruins of what had been Galveston's wharves. They saw the docks collapsed, boats wrecked. The greatest shipping port in Texas, one of the grandest in the nation, even in the world, had no wharf.

No living people could be seen on land. A light flickered from the darkness on shore, then disappeared.

Those on the sailboat who had family members in Galveston began leaping out of the boat in desperate anxiety. They splashed and climbed wildly ashore. In minutes, they had disappeared into the darkness on the island.

Monagan and the remaining crew stayed on the sailboat. Their plan was to wait till morning before setting foot on the island and entering the next stage of this awful journey. But as they anchored for the night, amid sights that made any sleep unlikely, they noticed something new.

A stench.

It drifted out over the bay. It was sickening, poisonous, impossible to ignore. The smell of putrefying corpses.

CHAPTER 12

THE PILE

Perched on their makeshift raft in the darkness of the night of horrors, the Cline family had been driven out to sea by nature's force. Then, at last, they felt the wind and tide change again. Soon they found themselves riding back toward Galveston. And as they approached the location of the island, they could tell the seas were less rough. The water level was lower. Debris in the air and the sea was less violent.

Some parts of the few remaining houses were visible. Here and there, some houses even had lights on.

Storms can't last forever, of course. Even the worst move on. They go and wreak their havoc somewhere else.

The Cline house had turned over at about 8:00 P.M. Now it was nearly midnight. The whole day had felt like an eternity, but the past four hours especially had seemed to occur somewhere unworldly, somewhere outside of normal time. The hurricane, Isaac and Joseph knew, was finally leaving Galveston, Texas.

Their raft ran up against a solid structure.

This seemed to be the ruin of a house—but it felt solid. The raft hung up on the building and got stuck.

There seemed to be people alive inside the house. The Clines and the little girl they'd rescued banged on the walls for help.

From inside this house, people began opening windows. People let them in.

Daisy Thorne could hardly comprehend that she was still alive. The sun rose over Galveston on the morning of Sunday, September 9, 1900, to reveal a day that promised to be one of the clearest and calmest of the summer. The humidity had broken. The sky was blue.

Daisy's neighbors the McCauleys had fallen with Mrs. Thorne's room into the sea. And yet in the Lucas Terrace building, one room remained standing. Only one room: it teetered on some wreckage of the ground floor, but it was semierect. The rest of the building had long since collapsed, but in this last room—Daisy's bedroom—were packed Mrs. Thorne, Daisy's sister, brother, and aunt, and nearly the whole crowd of neighbors who had sought refuge there. They had ridden out the climax in what amounted to a rickety, nearly toppled tower.

When they could tell it was over, they'd climbed down in the dark from their precarious perch, soaked and chilled.

Now they were all sitting outside the building. The thunder and lightning and slashing rain had stopped. Even one of Daisy's cats had survived, hiding in a drawer in that room.

By first light, the wind had died, and the gulf water, which had risen past the second story, had receded. The beach was revealed again, and the gulf itself lay beautifully calm in the morning sun.

But Daisy, her family, and their neighbors confronted a sight as shocking as anything they'd endured overnight. The intimate horrors of the storm experience were over. Now they faced new horrors, on an impossible scale.

As Daisy and her family and neighbors, huddled on wreckage, looked around at their city, what they saw seemed fantastical, something out of epic biblical prophecy, out of nightmare. Yet it was real. There seemed to be no city left at all.

In every direction, from the gulf to the bay, and as far east and west as the eye could see, the city of Galveston had been knocked down. Where a house or office building did remain standing, it stood out starkly against the bright blue sky. Even many of those hulking shapes turned out to be nothing but vertical wrecks.

The wreckage had created a new landscape. Piles of deeply, chaotically stacked debris, formerly buildings, along with everything formerly in those buildings, made hills, knolls, and mountains of randomly strewn beams, timbers, windows, lath, and brick, giving way to flattened plains of further wreckage, with personal items strewn about them too: toys, beds and bedding, dinnerware, tables and chairs.

Uprooted trees were in the wreckage, twisted poles and tracks, machinery. Galveston was sheer mess. It stretched on and on. It piled up high.

As she looked around in bewilderment, Daisy could feel soaked lime from the walls stinging her skin. Crumbled mortar and plaster clogged her red hair. She'd lost her shoes.

Other survivors began emerging that morning, all over town, from wherever they'd been lucky enough to e—
Nobody felt lucky. Stunned, awestruck with grief
unshakable bewilderment, people couldn't even get

ings. They couldn't tell, at first glance, uptown from downtown. People cast desperately about for some familiar landmark, some way to know which way was which and where they had ended up.

Nothing was recognizable. Overnight, the Galvestonians' physical world had become horribly flattened, a weird landscape of endless wreckage where their great city had stood.

What happened to cause this particularly gruesome effect, not only on the lives of so many dead Galvestonians and their loved ones, but also on the city itself, had largely to do with the relationship between two factors. One was the surge that occurred when the cycling wind finally got behind the tide. It sent a massive wave farther and higher into town than anyone had ever seen before.

The second factor, interacting with the first, was the island's high ground, around Broadway. For as that surge rolled quickly northward across the island out of the gulf with monumental force, it pushed many tons of debris before it—all the wreckage that the tide and the wind and rain had brought down on the low ground. That massive wreckage flowed uptown toward Broadway, and then began to pile up there, rising against the spine of land that had made Broadway seem the safest place to run.

The wreckage became a hard dam of wood and iron and shingle, stacked on top of the high ground. Behind that levee of ruination, the water that created it rose to heights it could not otherwise have reached. Then, pushed by the gulf wave behind it, the sea poured over the top of the dam. The water inundated the high ground, putting it under a flood.

The gulf used the ruins of the city's buildings to destroy the rest of the city. When the water receded, it left behind a new world of deep debris.

And it deposited human bodies in the debris—many times more of them than even the wildest rumors reaching Houston had foretold. To the survivors, the grief and shock that first morning were overwhelming. Anyone alive in this terrifying landscape of ruination was also someone terribly bereaved. Evidence of death was as palpable as could be.

Lying about the vast desert of wreckage was a multitude of corpses. These bodies, hard to take in, brought back all the horrors of the night. Almost everybody had expected to die. That some were alive and others killed, their bodies tossed upon these piles of debris, represented the height of randomness, sheer chance.

On that first morning alone, survivors saw many things they would never forget. Most of the corpses were stark naked: the water had torn their clothes from them. Some had died in positions of beseeching prayer. Dead mothers lay grasping their dead babies and children; other children's bodies lay alone, tossed.

Flying debris and falling walls and roofs had ripped people completely apart. Severed body parts were all over the piles.

There were bodies in vacant lots, bodies in standing pools, bodies piled up and bodies broken. Bodies hung in groups from broken railroad bridges.

And everything, human and inanimate alike, was soaking wet and covered with a weird slime, as thick and stubborn as grease.

Then horror surged higher. Survivors began to realize they were hearing muffled cries. The cries came from deep within the piles of debris. Buildings had fallen, so there were even more bodies under the wreckage than could be seen on top. And some of those bodies were still alive.

The survivors tried, on that first, desperate morning, to dig with their bare hands. They held out a hope of freeing those buried alive.

But the wreckage was too heavy and wet, the survivors too tired, weak, and not organized for work. The debris was fifteen feet deep in some places. There was nothing to do for the dead, and nothing to do for those still dying.

As the first morning brightened with the sun, new problems began to take immediate, pressing form. Galveston was cut off, the dazed survivors began to realize. They were alive, but the bridges and wharves were gone, the boats smashed, the cable and phone lines down.

There was no electric light. There was little food. There wasn't much drinking water: the pumping station that brought fresh water in from the mainland was shattered.

There was no way to escape the island, no way of letting help get in. Those who had survived the night could not, without some immense effort, survive much longer. And nobody in Galveston knew whether anyone outside Galveston was coming to help.

If little Louise Bristol and her family had been brave enough to leap into the water with the mattress and try swimming across to the neighbor's house, they might all have died. Then again, by staying in the room, they'd risked death too. Still, when the water started receding at about 11:00 on Saturday night, they were all alive.

And thanks to Cassie Bristol's forethought and energy, they had food, water, and even light upstairs.

Then, as they sat there on the morning of the ninth, with the floods gone and nothing but wreckage all around them, they heard the back wall of their own house fall. They went out on the porch to look at the damage.

Everything that had once been on the rear side of the Bristol house now lay in the yard next door: the stove, the contents of two bedrooms, the dining room furniture, all ruined. The entire back wall was down.

This house was Cassie's living. The ruined things in the yard represented everything she'd worked for in a ceaseless effort to keep her children out of poverty, proud and genteel.

One look told Cassie Bristol she was going to have to start all over again. New debts. Many more years of constant work against bad odds.

"Oh God," Louise heard her mother say. The despair and bitterness she heard filled the little girl with sorrow.

"Why couldn't we have all gone with it?" Cassie Bristol asked.

As the first morning after the storm went on, all of the urgent problems that kept arising were pushed aside by something even more urgent. The smell.

By now, the morning sun was directly hitting the wreckage. The day was getting warmer. The bodies were beginning to putrefy. And that meant rampaging disease and death were not far behind.

Worse: most of the rotting bodies were deep under the pile, seemingly immovable. New fears began to rage in Galveston.

Sorrow could not become the main thing. Along with somehow treating the sick and injured, and somehow housing the homeless, and somehow getting food and fresh water, the people of Galveston now faced the immediate task of somehow removing all of these corpses—both the ones on top of the pile and those buried under it—from the island.

If that effort should fail, further deaths would quickly start to ensue. There was no choice about this sickening task. There was no quarter. There was no time.

Nobody could say yet how high the death toll would be. The visible bodies alone were everywhere in their grotesque arrangements; many more were invisible, festering below the wreckage.

Galveston couldn't wait for help. Removing bodies required digging out right away. This project needed to be organized, it needed to be led. There was no time to mourn.

Leadership in Galveston took strange new forms. First thing Sunday morning, Mayor Walter Jones sent messengers to important citizens, calling an emergency meeting at the city's best hotel, the Tremont. That building, though badly damaged by water, was standing.

The meeting brought together the city's best-known and, in many cases, richest citizens. The Deep Water Committee had continued to exist, and to influence policy, long after the channels had been dredged. Some of its key members attended. Ike Kempner, son of the late Harris Kempner, a founder of the committee, was there. So was the committee founder John Sealy. Colonel J. H. Hawley, one of the best respected men in town, began playing an important role.

Police Chief Ed Ketchum hadn't left his post at City Hall all night. City Hall was partly in ruins, but after sending his son Henry home late Saturday afternoon—with the terse "it's going to get rough"—the chief hadn't seen his family, didn't yet know whether they'd survived.

His attitude was that if he was good to others, God would be good to him. The chief left City Hall for the Tremont.

Religious leaders came too, including the Catholic priest Father Kirwin, rector of St. Mary's Cathedral. The basilica was one of the buildings to survive the storm. Father Kirwin's good friend Rabbi Henry Cohen was there. Despite falling plaster, the Rabbi and Mollie and the children had made it through, and Cohen was already thinking of nothing but how to provide relief to his fellow citizens.

It took some doing for the city's surviving elite even to get to the Tremont. They had to pick their way through the corpses over sharp mountains of deadly wreckage. But convene they did.

When the team had gathered in the hotel lobby, Mayor Jones opened the meeting. He noted formally that no department or agency of the city's government was now functioning. He therefore formally invoked emergency executive powers.

The mayor created a body he called the Central Relief Committee. He bestowed on this committee unlimited powers to do anything it believed necessary to cope with the crisis. The committee, it was quickly decided, would answer only to a triumvirate: Colonel Hawley, Police Chief Ed Ketchum, and Mayor Jones himself. Under those auspices, it would run Galveston until the emergency could be declared over. With Galveston cut off from the rest of the world, the committee became its new government.

The mayor also declared martial law.

Disturbing rumors were already spreading. There was wild

drunkenness out in the street, people crazed by the horror of
their surroundings. Other stories told of "ghouls": looters already pillaging corpses for jewelry and other valuables, explor-
ing ruined buildings in search of money and property.

Also, some storekeepers with supplies that hadn't been
ruined were engaging in another form of looting: price gouging. Bacon, it was said, was up to fifty cents a pound this
morning. Bread was sixty cents a loaf. Captains and owners of
the few working boats were already meeting to fix prices for
trips to the mainland.

The mayor and the committee decided that the police and
the local militia must put an end to all of these practices immediately.

Local militia: that sounded good. But how to organize an ad
hoc military and police force under these impossible conditions? Who should run it?

Chief Ketchum said he had very few police officers alive
and functioning. The difficulty in building a police and militia force to exert some discipline on the city looked overwhelming.

But another man had shown up at this meeting at the
Tremont: one Lloyd R. D. Fayling. Fayling was a veteran soldier and officer. He'd served as a deputy U.S. marshal in Chicago during riots there; he'd led a company of volunteers in the
Spanish-American War. Now a civilian employed by a publishing company, Lloyd Fayling was a lover of sharp military
discipline.

He was also evidently fearless. During the horrible night
before, Fayling had saved more than forty people from drowning. Wearing only his trademark brightly colored bathing
suit—it had made him famous that summer on Galveston's

beaches—he'd helped occupants of a runaway sloop get into the Gill and League Building at Twenty-First and Market Streets, where his office was. Men had literally thrown babies from the boat into Fayling's arms. Then Fayling had used that same craft to make repeated excursions into the water. He'd rescued many others that way.

This morning, with the water mostly gone, Fayling had changed out of his bathing suit into something presentable. He'd climbed over piles of sharp glass and brick to get to the Tremont Hotel to offer his talents to the city. It was Fayling who brought in the news that people were wandering crazily in the streets. Some of them, he reported, were soldiers from Fort San Jacinto and Fort Crockett. Garrisons at both forts had suffered much loss of life and terrible damage to armaments.

The soldiers in the city's wreckage were nearly naked and profoundly confused, Fayling reported—but they might be made useful in restoring order. Fayling also told the meeting that looting had begun.

He suggested that Chief Ketchum appoint him a police officer. Ketchum found a wet envelope. On it he scribbled an official commission making Fayling a sergeant. The chief, the mayor, and the committee as a whole assigned him the job of forming a militia.

Some warned him that the soldiers out in the wreckage weren't likely to take orders from a civilian, but Fayling was nothing if not confident. Damp commission in hand, he went out into the ruined streets in search of his new army.

The next thing the committee determined it had to do, right away on Sunday morning, was get word out to the mainland, to the state capital at Austin, and to Washington, D.C.

The city owned a twenty-foot steamer with a draft shallow

enough to handle the bay. A small group volunteered to go to the ruined wharves, determine whether that boat was functioning, and try taking it across the bay and then make their way to Houston.

On Sunday, the group did make it across the bay on that small, badly damaged steam launch. The first official delegation to get out of Galveston, they arrived on the coast at Texas City on Sunday afternoon.

As these men began wading across the flooded plain, they saw how far the devastation had reached. They passed dead bodies and debris. At last they found some functioning railroad tracks.

On the tracks was a handcar. They climbed on. Pumping the handle up and down, they rode the rails toward Houston to bring the awful news and ask for help.

About four blocks from the Tremont, Lloyd Fayling, newly minted police sergeant with extraordinary emergency powers, came across four artillery soldiers. They were barefoot. They seemed bewildered and lost.

" 'ten-SHUN!" Fayling snapped. As if automatically, the men responded. They came to attention and awaited orders. Fayling had the impression that they were relieved to have a leader.

They were now policemen, Fayling informed these four men. He would swear them in later, if he had time, but for now they were to fall in and follow him in the task of locating clothes, guns, and ammunition.

When two of the men began complaining and asking questions, Fayling barked, "Silence in the ranks!"

The men became quiet. Now Fayling saw an army captain he knew and respected, standing on a ruined corner, wearing pajamas. Fayling verbally commissioned this captain an officer of the new police force.

With the pajama'd captain's help, he led his recruits slowly through the wreckage in search of shoes, guns, clubs, and food for the small company. On every corner, they found new soldiers and tried to outfit them as well.

Then they found a bugler militiaman who had brought his instrument out into the wreckage. Fayling ordered him to blow the call for militia assembly.

The bugler blew as loud as he could. And amazingly, the Galveston militia began to assemble. Climbing over wreckage from every direction emerged men ready to serve. Many had even changed into their militia uniforms. Fayling began to think there was hope for civic order after all.

Only two hours after he'd been appointed a police sergeant by Chief Ketchum at the Tremont, Lloyd Fayling had armed men standing guard duty all over the former downtown. Pleased, he went back to the Tremont.

He wanted to report to the committee. And he wanted more power.

To dispose of the thousands of corpses, the committee first had to dig out and find the bodies before infestation set in. That meant cleaning up the wreckage, starting right away.

The committee had given itself extraordinary powers, and it had declared martial law. Now it used that authority.

The committee established a makeshift morgue in a cavernous warehouse still standing on the Strand between Twenty-First and Twenty-Second Streets. The dead bodies were to be brought there and laid out in long rows. Distraught families soon began arriving in hopes of identifying loved ones.

But as the day went on, the warehouse got hot and began

to stink. There were only about fifty bodies in the warehouse morgue so far. There were untold others on and under the pile.

This morgue idea was dwarfed by the gigantic scope of the grotesque problem the survivors were facing. It wouldn't work.

Sergeant Lloyd Fayling arrived back at the Tremont Hotel late Sunday morning. He reported to the committee on the new emergency force he'd begun building out of the shards of the army and the militia. Mayor Jones and Chief Ketchum, deeply involved in planning, quickly gave Fayling what he'd come for: a new commission.

Now he was Major Fayling—commander in chief of all military and police in Galveston. He had power, granted by the committee, to draft any men he met to serve under him and to requisition any property he deemed useful.

He was subordinate only to the chairman of the committee, the mayor, and the police chief. Armed with these new powers, Major Fayling went back out into the ruined streets.

The Dallas insurance man Thomas Monagan had spent a horrible night anchored on the bay amid the stench. On Monday morning, he had the schooner's captain weigh anchor. They sailed over to what remained of the wharf.

There they left the boat. They entered the ruined city—the first official civilian relief delegation to Galveston.

U.S. General McKibben and his men had preceded them in the small steamer, along with Texas General Scurry, as the first relief delegation of the U.S. and Texas military. McKibben's trip across the bay had shaken the military man as hard as Monagan's trip had shaken the insurance man. After steering around the floating bodies of women and children, "I have not slept a single moment," McKibben told people days later.

And yet these first outside relief teams also found that the people of Galveston hadn't been waiting helplessly and hopelessly cowering until relief arrived. Anything but. It was only Monday morning—the beginning of the second day after the storm—when Thomas Monagan and his crew entered the city. To their surprise, they found a form of officialdom in action.

The committee had taken charge. A stunned people and a group of overwhelmed civic leaders had already started responding to this nightmarish scene. Thomas Monagan felt an unexpected sense of the city's resilience and vitality in the midst of awful tragedy.

Much of what Monagan found to admire in the ruined city reflected the energy of Major Lloyd Fayling. His first exertion of martial law on Sunday morning had been the closing of all saloons, under orders of Chief Ketchum. Given his powers, there was little practical resistance. Fayling ordered his men to tell all saloonkeepers to close up shop. If they encountered any complaints, they were to make a show of arms. If anyone reopened, they would arrest him.

The other big issue was looting. Fayling gave orders: any looter was to be shot on sight, no questions asked. Next he would turn his attention to price gouging.

But it was becoming clear to Fayling that his force needed to be bigger. Much bigger. Putting out a call for volunteers, he also began using his powers to impress troops at gunpoint.

Fayling soon had three companies of infantry. These were named A, B, and C—with the A Company made up of regular soldiers, and the B and C Companies formed by mixing militia with impressed and volunteer civilians.

Then Fayling got a cavalry troop. In the first two days after the storm, having stationed infantrymen at key points, he'd been patrolling the entire city on foot with his personal squad. They ranged widely, in part to extend patrols into outlying areas, in part to supervise the many squads of infantry stationed throughout the area.

But the piles of wreckage were sharp, with broken glass and tile, nails, and pointed lumber. Fayling and his men were ruining their shoes and slicing up their feet. At the armory in the mornings, the men had to bathe their feet in cold water just to get into their battered shoes. They walked for miles when, Fayling knew, they really should have been in a hospital.

A solution lay in yet another weird, post-hurricane phenomenon. All over town wandered surviving domestic animals and livestock: cows, dogs, cats, and horses. They'd been abandoned. And the horses, in particular, eagerly sought out human company.

These were Texas horses. Some were adept at galloping and jumping under tough riders—they'd been bred to that life. Many others were not: they were work animals, dray horses.

But Fayling's men were Texas riders. They knew how to train horses. Very soon, Fayling and his mounted troops were flying about town at full gallop. The horses were leaping the deadly wreckage, clearing piles five feet high that were otherwise impossible to pass.

With this cavalry division, Fayling now seemed to be everywhere. He was going through horses the way he'd gone through shoes: he wore out two a day. His plan was to never stop moving. He didn't sleep. He ate sandwiches in the saddle.

Nobody on sentry duty could relax for a moment: Major Fayling could seemingly materialize out of thin air in any part of town at any time. Indeed, he made sure the men believed that he would summarily shoot anyone caught sleeping on guard duty. He wouldn't have done it, he said later. But he wanted the men to think he would, and they did. He never caught any sentinel napping.

Fayling also requisitioned every firearm he could find in Galveston. (He personally carried two six-shooters and a saber, which he used for making arrests.) Mayor Jones issued an order under martial law putting the citizenry on notice that all arms must be turned over to the military. All citizens were forbidden to carry arms without written permission from Mayor Jones, Chief Ketchum, or Major Fayling.

Prices, meanwhile, were set by the committee and enforced by Fayling. Soon bacon was down to fifteen cents a pound, bread to ten cents a loaf. And a trip across the bay to the mainland—they were beginning to happen—was $1.50 per passenger.

And yet even before the first relief teams arrived from Houston, Austin, and Washington, and despite their own staunch early efforts to improve a disastrous situation, the committee and all the people of Galveston had quickly come to realize that the horrible smell of decay could only get worse. Many people seemed to have entered a state of deadly calm and total focus. A normal approach to burying the dead—identifying loved ones and giving them decent rites—was clearly out of the question now.

As early as Monday morning, the committee had therefore arrived at a new solution. These bodies must be thrown into the sea. Right away. Stop laying bodies out for a decent burial that could never take place: mass consignment to a watery grave was to be indiscriminate, total. There was no choice.

And so the loading began. All day Monday, every possible

conveyance was pressed into service for hauling bodies: carts, drays, wagons, and fire trucks.

Fleming Smizer, father of the newlywed Annie McCullough, worked at the Custom House at Sabine Pass across the bay on the mainland, and now he was crazy with worry. He'd weathered the storm but still had no news of his wife and daughter.

He didn't know they'd sought shelter in the school on the high ground of Broadway. He couldn't know whether any building where they might have sheltered had withstood the storm. He only knew there might be thousands dead in Galveston.

Smizer couldn't get across the bay without a tugboat, so he was forced to wait as his tension increased. Finally he did board a tug for Galveston.

On the island at last, Smizer found the school at Broadway and Tenth nothing but a pile of wreckage. There Annie and Ed and their families had fled for safety, along with so many others. Broadway was the high ground, the thinking went; the waters would not reach it. But that high levee of timber and other objects shoved relentlessly from the gulf beach against the high ground on Broadway had brought about the ruin of the school, seemingly the sturdiest of buildings. In the end, it went down.

Smizer learned that homeless people were gathered now at the county courthouse—basically camping. He hurried there, and he was overjoyed to find his daughter, his wife, and his son-in-law among the crowd in the large building. They were unhurt.

And it was amazing that they were unhurt. At the school, Annie McCullough had seen fifteen people killed in one shot when lightning knocked a chimney down on top of them, right where she'd just been sitting. There had been a little boy—a white child—hanging onto Annie, begging her and Ed to look out the window and see if his house was still standing. The boy's father had gone back to find his wife. The man had never returned. With most of the school building wrecked, people inside had begun to face the fact that the last rooms must soon go too.

But then, around 9:00 P.M., Annie saw the moon. She could tell the waters were receding at last.

"Come," Ed told her. "Let's go down to the courthouse." He believed that building would still be standing, and he was right. He and Annie knew that not only Annie's roses, so carefully placed in tubs, but also their entire house, would be gone now.

So the couple and their families left the ruined building. Moonlight revealed the devastation. They picked their way through the slime and sharp edges of a horribly fallen city.

Annie, Ed, and the other Smizers and McCulloughs moved into the courthouse. There was nothing else to do. When Annie's father found them there, the whole extended family was reunited. Yet many of their friends had died. Annie and Ed themselves had no home and nothing left. With their parents and aunts and cousins, these newlyweds were now trying to live in a courthouse.

There was a white man at the courthouse where Annie and Ed and the McCullough and Smizer families were sheltering, and he kept breaking down in tears at the plight of the homeless people there. "If your house is gone," the man told Ed McCullough, "you bring your family. Come live with us."

Curious, Ed went over to the man's house. He found the place a muddy mess, but it was still standing. The man's wife was down on her knees, trying to scrape up mud. The man told her to turn the house over to the McCulloughs. Ed went back to the courthouse and retrieved his relatives.

They all moved into the white family's house. They began fixing the house up, and they would live there until they could get to Annie's mother's family on the Texas mainland.

Annie would always remember that white man. She would always recall the bright spot he represented in what, amid disaster, were becoming newly strained race relations in Galveston.

And yet Lloyd Fayling's men came for Fleming Smizer. The troops arrived where Ed, Annie, and the family were staying. They began searching the house for able-bodied men.

Ed was out at that moment. The armed men forced Annie's father out of the house. He had no choice but to go with them.

"I'm a government man," Smizer protested. "I worked in the Custom House." But Galveston's African American men were being rounded up and pressed into service.

Beginning with the horrible task of burying the putrefying corpses at sea, Major Fayling's troops began going around town, with guns pointed, stopping black men at random and demanding that they come down to the wharf and begin loading corpses onto a large barge. If the men objected, Fayling's order was "Load with ball cartridge, take aim!" Rifles were raised and aimed, and not surprisingly, compliance in the sickening task was total.

This singling out of men that white Galvestonians routinely referred to in public as "Negroes" marked another eerie effect of the hurricane. There had, of course, never been racial equality in the city. While African Americans had long played important roles in civic life, while the docks had seen the progress of the Negro Longshoreman's Union, and while all kinds of people mingled in the city streets, much of Galveston's social and political life had long involved rigid racial segregation. And the most menial and unpleasant kinds of labor had long been presumed to be a natural province of black men and women.

The prevailing white idea in 1900 was that "the Negro" was naturally less affected by the rigors of hard labor; less sensitive to assaults of bad smells and disgusting textures; and inured, through long generations, to submission. That philosophy prevailed more or less openly, not only in Galveston but throughout Texas and the American South, and indeed throughout the rest of the booming country as well. Even under normal circumstances, the most advanced and polite white Protestants routinely referred to anyone non-white and non-Protestant by race and religion first. Somebody would be called "a Negro," "a Jew," "a Greek," "a Roman Catholic" before anything else. Only a fellow white Protestant did not require such definition.

Even when approving of black people's or Jews' behavior, the terms were patronizing. Major Fayling, having stationed his African American servant Ed Hearde with a rifle to guard a group of thieves—who were also black—called Hearde "the bravest and most faithful Negro I have ever known." Weeks later, Fayling was in Houston arranging further relief, and he was met officially by Louis J. Tuffy, mayor pro-tem of that city. Fayling referred to Tuffy not by name or by title but as "a fleshy Hebrew gentleman."

Major Fayling's snarky bigotry may have been extreme, even for his day. Rabbi Henry Cohen was serving, without objection and with utter respect, on the Central Relief Committee, as was the Catholic Father Kirwin. Jews had high standing in Galveston's elite, and a wealthy Mexican, Thomas Gonzales, had been among the city's leading citizens.

The hardest lines, though, had always been drawn between those in the city who defined themselves as white, and the descendants of those who had been enslaved. Now, in the immediate aftermath of the storm, with martial law declared in Galveston and troops dispatched to enforce it, something even grimmer was turning up in the city's racial culture.

With death everywhere, both visible and in the stench of the air itself, new fears arose among many white citizens, fears of a violent and criminal element, whom they took to be black, not white.

The stories had begun flying on the first day. Those "ghouls" who were looting the dead bodies on the pile and robbing the empty buildings? They were Negroes, white people reported. So savage were these black ghouls, it was said, that they cut fingers and ears off corpses to get at rings.

Rumor topped rumor: the ghouls actually chewed dead fingers off by the handful and filled their pockets with them. "The Negro," though supposedly used to submission, was also somehow apt to become bizarrely and inhumanly savage.

Looting did in fact occur. There may even have been lopping off of fingers. But that crime was a racially integrated activity. Major Fayling himself, along with many others, reported that fact. "I am sorry to say that white men are side by side [with the Negroes] in their damnable work," one white visitor described.

"The ghouls are composed of Negroes and foreigners," another man reported confidently.

So despite the obvious presence of white looters, many white people—including the nearly all-powerful Major Fayling and his troops—referred routinely to the looters as Negroes, joined, they were sometimes willing to admit, by less-desirable whites. The emergence in Galveston of a criminality so shockingly selfish, heedless, and low simply had to be associated, by many white citizens, with a race they considered inherently "other."

For Fleming Smizer, a descendent of one of Galveston's oldest families, so recently reunited with his wife and daughter, and for so many others, there would be no deferment and no quarter. White soldiers forced black men at gunpoint to the front lines of the most horrifying labor that any city could ever face.

That first Monday afternoon, citizens stood about the wreckage watching a horrible kind of parade. Carts and wagons kept arriving at the wharf on the bay side. They were piled with bodies.

All afternoon, fifty African American Galvestonians, sweating in the sun, fighting off their nausea, lifted about 700 dead bodies of all races and ages from those carts onto the barge. The men threw the corpses into the hold at random. People watched from the waterside, stunned and fascinated.

Major Fayling did feel bad for the workers on the barge. He knew this job was an awful one. He ordered whiskey served to the men throughout the day.

When the barge was fully loaded, it was attached to a tugboat. The tug towed the barge—with the corpses, the workers, and some of Fayling's troops—out of the bay and around the end of the island.

The tug steamed eighteen miles into the Gulf of Mexico. Here is where the committee had decided the bodies would go under.

But it was dark now—too late to complete the task. The

barge anchored. The living had to spend that night on the water with the dead.

When the sun rose on the gulf, the next phase of work began. Lifting the bodies out of the hold, the men lashed a weight to each one so that it would sink. They used a random assortment of weights: pieces of storm-broken iron, window-sash weights, anything heavy enough to send a body down and keep it there.

Then the men lifted the hundreds of weighted bodies. They dumped them all into the gulf. That was Tuesday morning.

Tuesday afternoon, the bodies came back. They arrived with the tide on Galveston's beaches. Soon corpses littered the sand.

The gulf had spoken. It denied Galveston's dead burial at sea. It denied the survivors any hope, however gruesome, for a quick end to their living nightmare.

CHAPTER 13

"I CAN BEGIN LIFE AGAIN, AS I ENTERED IT"

AMID ALL THIS FAILED CIVIC EFFORT TO COPE WITH AN UN-imaginable public horror, each surviving citizen in Galveston tried to sort things out. In those first days, people on the mainland who had loved ones in Galveston were still trying to get in; people in Galveston were trying to clean up, help others, get systems working. All of that took place under a pall of horrified amazement at what they were seeing, overwhelmed by total loss.

One of the things that everybody—whether stuck on the nightmarish island, or outside trying to get in—had to cope with was a daily listing of the dead. One of the earliest civic institutions to get up and running after the storm was the press. The *Galveston Daily News* was first, followed by the *Galves-*

ton Tribune. Late on that first stunning Sunday, the very first day after the storm, the *News* was published.

It was only a small single sheet. All it offered was a stark list: the names of the dead, as reported by grieving survivors.

For some time, lists were the main thing the *News* offered. And the list entitled "Dead" lengthened horribly day after day.

The body count was still provisional. Really, it always would be. But the first hopeful estimates—that only a few hundred people might have died on Saturday night—succumbed quickly to a staggering reality.

Five hundred had once seemed a hysterical, outsized exaggeration. It soon became clear that the death toll of the Galveston hurricane numbered in the many thousands. With decomposition setting in, there was no time or stomach for careful counting. The most conservative estimates would place the count at 5,000. Others would suggest it came near 15,000. Some split the difference, and around 10,000 seems a fair guess.

The *News* soon started running a second list, entitled "Not Dead." That list was unfortunately short but critically important. As people were found alive, the paper took their names off the "Dead" list and filed them under "Not Dead."

Nell Hertford, frequently escorted by the morose Boyer Gonzales, found that life in the ruined city had taken on an eerie calm.

To others, people seemed on the brink of madness. Not only were so many bodies being unearthed from the pile: new stories of that night of horrors kept coming up too. The terrible stories made the terror seem to go on and on.

Of the children who had sung for help to the Queen of the Waves, those orphans of St. Mary's, only three had survived.

All of the other orphans—along with Sister Elizabeth Ryan, Mother Camillus Tracy, and all of the nuns and other adults who cared for the children—died in the water.

The three survivors were boys in their early teens: Albert Campbell, Will Murney, and Francis Bolenick. Older kids, they probably hadn't been roped to the adults. Swept through the wreckage by the powerful currents, all three had managed to grab the same uprooted treetop.

The tree was stuck precariously in the gulf between the masts of a wrecked schooner. The boys clung to the treetop together and held their bodies against the raging tide for long hours.

At one point Albert, weakening, shouted that he was drowning. Will grabbed a piece of rope from the schooner wreckage. He tied Albert to the treetop, and they all kept fighting the water.

Then the schooner masts broke, and the tree was abruptly released to the sea. Still the trio managed to stay with the tree as it flowed back toward shore.

The next morning, the three orphans found themselves on the gulf beach. Littering the sand were the dead bodies of everyone they knew.

Strange stories of light pierced the darkness. Arnold Wolfram and the ten-year-old Western Union boy made it across that plank at the last minute and got onto Wolfram's friend's upper porch. Having fought their way through town together, the man and the boy survived together. The next day, Wolfram restored the kid to his grateful family and rejoined his own.

The Delzes—a family living on the west end, near Daisy Thorne's apartment house—were trying to come to grips with a terrible loss. Sixteen-year-old Anna Delz had been washed into the gulf by the storm.

And yet when Anna's father was finally able to communicate with his sister, who lived on the mainland, he learned that Anna was alive. She'd been swept from the gulf back onto the island, then across the entire island, and then out into the bay, and then all the way over to the mainland coast.

There the exhausted girl slept on a pile of lumber just above water level. At daybreak Sunday, she walked—naked, since the storm had taken her clothing—until she found a house and asked where she was. Learning she was in the town of La Marque, Texas, Anna found the aunt she knew lived there. Soon Anna was restored to her family in Galveston.

As the tales of Saturday night were told, the ruined island city took on a new and eerie strangeness. Open cremation fires, strategically sited among the wreckage, scorched through the days and lit up the nights.

Burial at sea had failed. The stench was more powerful every hour.

This was the Central Relief Committee's new plan. All bodies were be dug out of the pile and burned. Night and day, the dead would go up in huge bonfires until they were all gone. To the reek of putrefaction was added the stench of burning flesh. And because the burning went on day and night, a foul, ashy smoke now always hung thickly over the island.

The project was so massive that the crews who cleared the wreckage, turned up bodies, gathered them at the pyres, and threw them on the fire had to be massive too. While African Americans were always the first choice for such awful work, Major Fayling realized he had to stop discriminating so rigidly on the basis of race. At one point his troops roved the foot of Tremont Street, forcing every able-bodied man they saw into service at bayonet point. The major was on standing orders from the committee to shoot not only looters but also any man refusing to work on the "dead gangs," as people started calling these crews. He passed those orders on to his deputies.

Again, Fayling was careful to give the clear impression that he would gun down, without compunction, anybody refusing to obey orders. That attitude turned out to be enough.

Unlimited free whiskey remained the only way he knew of trying to ameliorate the physical and mental anguish of the dead gangs. Troops served booze in goblets to the workers. The men drank the whiskey steadily, but Fayling never saw any man drunk. They worked thirty-minute shifts, then took a break. Fayling knew that was all anyone could stand.

And so the labor went on, and a grisly rhythm took over. Digging was a daytime activity: crews went about the slow, tedious removal of tons of random wreckage—punctuated by abrupt, horrific discoveries of bodies and body parts, badly decaying now. Remains had to be handled, even as they decomposed in the workers' hands. Flesh was removed to carts and wheeled to the pyres.

Burning was a twenty-four-hour job. There were fires everywhere, all the time. It wasn't only smells and smoke that nobody in town could avoid: it was the sight of those bonfires, and of the bodies burning on them. These weren't crematorium furnaces: bodies burned with agonizing slowness on the pyres. Children of 1900 Galveston would grow old with those sights still vivid in their memories.

Little Louise Bristol wasn't allowed to go outside. Her older brothers were pressed into service, but through those long, strange weeks, Cassie kept the little girl inside the ruined boardinghouse. Louise smelled the stench but never saw the fires.

When the young lawyer Clarence Howth was thrown by the ocean waves, blasting through a window into his attic, far from his wife Marie and their newborn baby, his father-in-law and brother-in-law were swept away too. Clarence had tried, as he went underwater, to welcome his own end.

He hoped to meet his beloved wife in heaven. He opened his mouth to speed the drowning. But he couldn't do it. Trapped in turbulent water and rubble, his body fought for survival against his will. There was no sound. Fastened under his wrecked house, he began losing his unwilling fight to live.

And then he wasn't losing. The ruins released him; he swam to the top. Then Clarence, like so many others, was swept out to sea. For ten hours he fought waters that pushed him off down the beach to the southwest and then out into the gulf. His only float was a broken window frame.

When the waters receded, Clarence found himself back in town. He was badly bruised and cut up, shivering, naked, alive. He lay down on a pile of wreckage and cried.

And in the ensuing days, dazed by emotional turmoil, he knew that his wife and baby were getting piled into carts with thousands of others and slowly burned in pyres. For days he couldn't even walk. He had nothing left. No family. No home.

He wrote to his brother in New York. "All is gone!" Clarence told his brother in a howl of sheer despair. The despair was for both himself and all of Galveston. "Martial law, consternation, ruin, and starvation prevail," he reported.

And yet Clarence also reported that he was being well cared for. "My limbs are left and my mind only slightly impaired," he told his brother.

In the midst of deep anguish over total loss, something still stirred within the bereft young man. He summoned courage from somewhere. "I can begin life again," he said, "as I entered it."

Isaac and Joseph Cline, battered and bereft like many of their fellow citizens, spent the first few days after the storm trying to master grief and exhaustion. Isaac didn't pick his way through ruin to the Levy Building. Joseph went, but he couldn't get much done.

The Levy Building was still standing. After Joseph had left on that awful Saturday evening, the Weather Bureau office was manned only by the young assistant, John Blagden. And all that night, the building had rocked and rolled with the unbelievable winds.

Blagden held his post all night. The only instrument he had to work with was a barometer. The wind had snatched from the roof not only the storm-warning flag but also the anemometer and the rain gauge.

All night as the building quaked, Blagden watched the barometer fall and felt the wind pick up. With winds now blowing at speeds well in excess of what weathermen believed possible, the barometer plummeted, that evening, to depths that Blagden—along with the Clines, Director Moore, and all other American weather scientists—had never seen before.

The barometer stopped falling at 28.48. That reading became the bureau's new record low. Scientists later determined that the real barometric pressure in Galveston that night was 27.50, an even more shocking number. The real wind speed was later estimated at more than 120 miles per hour.

Blagden must have thought the Apocalypse had come. Nothing could hold out against a low-pressure zone like that. The wind meanwhile removed the entire fourth floor from the Moody Bank, just down the street from the Levy Building.

But John Blagden survived, along with the building. And while Galveston remained for a time cut off from the bureau in Washington, Blagden was able to return to work nearly as soon as the storm was over.

Both Cline brothers, however, found returning to work more difficult. Both were deeply shaken. Joseph had been injured during his escape with the children through the windows of the toppling house; his lymph glands had now become seriously swollen, and sometimes he couldn't get out of bed.

There was something deeper than physical injury, too. Joseph found himself emotionally shattered. He feared the effect might be permanent.

Still he went into the office and ran it as best he could alone. Under Joseph's direction, Blagden installed some temporary replacement instruments on the roof of the Levy Building. When telegraph service could be restored, the Galveston weather station office would be up and running and ready to report.

Isaac Cline didn't return to work for a week. Though he and Joseph had, nearly miraculously, saved all of his children, Isaac had lost Cora and their unborn child.

And he knew, as did so many others, that Cora's remains were somewhere in the pile. He expected they would be gathered up anonymously, carted, burned.

It will never be clear how far Isaac Cline blamed himself during those first grim days of burning pyres and martial law for the disaster that had taken the lives of so many people, including that of his own wife. Building a seawall would have lessened the damage and saved many lives. That wall hadn't been built, and it was Isaac Cline, chief U.S. meteorologist for Galveston and the whole Texas section, who had advised the city in no uncertain terms that constructing a seawall was entirely unnecessary. He'd called anyone fearing a hurricane in Galveston delusional.

Such stark, overpowering evidence of his personal failure—and the failure of his knowledge, and that of his colleagues and superiors—lay all about him as far as the eye could see. His own dear wife was one of the victims of that failure. Nearly all of the fifty neighbors who had sought shelter in Cline's home had died too. Isaac Cline's professional certainty of the nature of hurricanes had caused pain that was entirely personal.

Yet Cline continued to believe he'd done everything he could in the fulfillment of his duty. And it's possible, given the beliefs of his day, and the orders of his chief—not to mention his chief's blackout of all weather news from Cuba—that Cline had indeed done his best.

By going around town on Saturday, warning people and urging them to evacuate, he directly violated Director Moore's sole authority to issue such warnings. Cline felt the violation was necessary to save lives, and he believed it had in fact saved many. Without those warnings, he estimated the loss of life at twice what it was in fact.

That may have been Cline's wishful thinking. Even the claim that he gave personal warnings on the beach has been questioned. But he certainly did send the telegram to Moore in Washington, noting the presence, never seen before, of a high tide against the opposing wind. Cline himself, Dr. Young, and others took that condition as the sign of a potentially devastating storm.

Joseph Cline, whose accounts of those days do not always perfectly match his brother's, also said later that Isaac went to the beach to give warnings while Joseph went to the office. Isaac's phone call late that stormy day, in which the elder brother dictated to the younger the contents of a final, desperate telegram to Washington, came from somewhere near the beach, according to Joseph.

Director Moore, for his part, never publicly criticized Isaac

Cline for any action taken in advance of the storm. Moore believed Cline had been right to breach protocol, to take it upon himself to make warnings. The director never suggested that the Galveston weatherman had been either too lax or too aggressive.

Later, Moore even gave Isaac Cline credit for braving the tides and winds to phone in that final telegram from Galveston to Washington. Isaac made sure to correct the record. He placed credit for that feat where it belonged: with his brother Joseph.

Another miracle came with the little girl that the Cline family had rescued from the water that night and carried on their makeshift set of debris-rafts. In the morning, they'd left her with the family who lived in the home where they'd at last found shelter. They assured the girl they would find a way to provide for her.

The girl told them she lived in San Antonio and had been visiting her grandparents in Galveston with her mother. In a resonant coincidence, the girl had the name Cora—the name of Isaac's lost wife. They wrote it down. They hoped to locate the grandparents.

A few days later, Joseph was in a drugstore, seeking relief for his shaky nerves and swollen glands. He heard a grief-stricken man describing himself to the druggist as coming from San Antonio. Joseph had a feeling. He asked the man if he knew the child.

The man turned to Joseph a face deeply lined with sorrow. "She is my daughter," he said.

Joseph soon brought father and daughter together.

Others on the mainland with loved ones in Galveston were still desperate to get in. Dr. Joe Gilbert, engaged to the schoolteacher Daisy Thorne, was in Austin when he heard to his shock on Sunday about the Galveston disaster. He took the next train to Houston.

Joe had to get into Galveston. He had to find Daisy.

In Houston, Joe grabbed the first newspaper he saw and anxiously skimmed the list of the Galveston dead. In a quick, horrible moment his worst fears came alive. Daisy was on the list, with all of her family. Joe reeled. His life hit bottom.

But Joe was saved from despair. Father Kirwin had arrived in Houston, and when Joe, stunned, ran into the priest, Kirwin told him the paper was wrong.

Daisy was alive, miraculously enough, and so was the whole Thorne family. Kirwin had seen them.

And it was on Monday morning, while still in Houston, that Dr. Joe began realizing with joy and awe just how miraculous Daisy's survival really was. The full news had started coming in from Galveston now. The papers were packed with details of horror, both rumor and truth.

Joe boarded a train heading south for the coast. On the car he found an old friend, also a doctor, traveling home to Galveston in hopes of lending medical help. As they rode, the friend asked Joe what he planned to do in Galveston.

"Find Daisy," Joe replied.

"And marry her," he added.

The aspiring painter Boyer Gonzales, reluctant scion of the family business in Galveston, was outside Galveston too, having spent part of the summer studying color theory with Lansil in Boston. Then he'd traveled to Prouts Neck, Maine, for some idyllic weeks of work with his mentor and friend, the famous artist Winslow Homer. That's what Boyer loved to do.

And yet he'd begun this turn-of-the-century year in a state

of desolation and loneliness. His relationship with the family business was based on a bleak sense of obligation to carry on something for which he had no love or knack. It meant sacrificing his ardent desire to grow and bloom as a painter.

His bad stomach had worsened. He'd made one of his now-annual trips to Kellogg's sanitarium in Battle Creek, Michigan.

So at the end of summer, Boyer Gonzales remained in a state of conflict over his own ambitions and his obligation to Galveston and his family legacy there. It left him still emotionally paralyzed, distant, and in physical pain. And in his friendship with his frequent companion Nell Herford, he remained terminally noncommittal.

In the first week of September, Boyer said good-bye to Winslow Homer and Prouts Neck and traveled unhappily back to Boston. He was facing the grim prospect of a return to work he increasingly detested.

In Boston on Sunday, September 9, he received the shocking news that a monstrous hurricane had devastated his hometown, and all communication was out. Boyer wrote immediately from Boston to his younger brother, Alcie, who had been living in the Gonzales mansion on Avenue O.

Boyer did not, characteristically enough, reach out to Nell. He did write to Nell's brother-in-law Walter Beadles, with whom Nell lived, and with whom the Gonzalez family had long done business in the cotton trade.

Nell, however, took it upon herself to answer his letter to Walter. She wanted to fill him in personally on the horrible conditions in town.

"My dear Mr. Gonzales," Nell began her letter. She was addressing her escort in terms at once notably affectionate and perfectly formal. Such were the terms of their relationship.

Nell let Boyer know that the Gonzales house had been badly damaged. She reassured him that Alcie, she, and many others he knew had survived without injury.

Nell had always wanted to keep letters to her depressive friend cheery, upbeat, newsy. This one had to be, as she put it, "gruesome." She said she would now gladly turn her back on Galveston forever.

And Nell closed her letter by noting that if she were selfish, she'd want Mr. Gonzales to be there in Galveston. "But for your own sake," she wrote, "I advise you not to come home if you can possibly help it."

But Boyer did come home.

"I can begin life again, as I entered it."

That's what the devastated young lawyer Clarence Howth said. That's what he was trying to believe.

Over the coming weeks, some people would show amazing guts and resilience. While some seemed on the verge of madness, others seemed able to consider rebuilding their ruined lives from nothing.

They seemed capable of making new plans.

But could the city of Galveston begin life again? This great, thriving Texan jewel in the gulf, where years earlier Jean Lafitte had planted an outlaw settlement, had gained national and even global prominence at dizzying speed. The New York of the Southwest had been on track to take its place as one of the first cities of the booming American nation, even as that nation assumed a leading role on the world stage in the twentieth century.

All of that was gone. The busy wharves smashed, million-aires' mansions and ordinary homes tumbled, the churches and banks and office buildings in ruins. At least a third of the city was demolished: more than 2,600 buildings entirely destroyed; an astronomical $20 million or more in property loss; a shipping fleet disabled.

And martial law reigned. Government was in the hands of an ad hoc committee whose most important job was to get rotting flesh pulled out of miles of slime-covered wreckage, then burned day and night in pyres.

To many—if they even looked up to consider matters beyond the grim tasks at hand—any prospect for Galveston's future looked unutterably bleak. Many people were planning to leave town by boat as soon as communications were restored with the mainland. Nell Hertford, writing Boyer Gonzales, wasn't alone in declaring herself ready to gladly turn her back on Galveston forever. She'd advised Boyer to stay away from home as long as possible.

But would that desire to leave mean temporary withdrawal? Would people ever return to the city, ready to rebuild it, more or less from scratch?

Or were the people of Galveston, once so cavalier, about to follow the example of those who had abandoned Indianola after the last great hurricane? Galvestonians now understood, in the most horrific terms, what a hurricane could really do to human society on their island. They knew Galveston wasn't specially spared from disaster. They knew, too late.

And even if the city could be rebuilt, who would ever invest any money there? How could such a city thrive?

Galveston's identity, its very existence, was tied up in its excellence as a port. The great rail and freight and cotton and shipping companies had suffered extraordinary property losses in this storm. Those companies' directors would now know what a single storm could do to their investments.

Rebuilding a great city on a sandbar, now proven vulnerable to devastating natural fury? That struck many as the height of foolishness.

In the meantime, Galveston, once so proud, even sometimes superior, was losing its brave battle for mere survival. The committee and the citizens never stopped working. The city of Houston and the U.S. Army were doing their bit. But hunger, thirst, disease, grief, and isolation would win, unless Galveston got more help.

CHAPTER 14

"IN PITY'S NAME, IN AMERICA'S NAME"

WILLIAM RANDOLPH HEARST WAS AT WAR WITH JOSEPH PUlitzer. In 1900, both men were publishers with major newspapers in New York City. But Hearst had long been the upstart, and Pulitzer was still the man he had to beat.

Even when he'd begun, back in San Francisco, as the young publisher of the *San Francisco Examiner,* William Hearst had placed the elder, better-established competitor in New York City in his sights. It was Pulitzer, publisher of the *New York World,* who had pioneered the sensational pop-news style that Hearst brought to a high pitch of hysteria in the *Examiner.* Highbrows sneered at these cheap, popular news organs, "the yellow press," but everybody wanted to read them.

And like Hearst, Pulitzer had originally begun outside New York; he'd made the *Post-Dispatch* the most popular paper in

St. Louis. Then in 1883 he bought the *World,* moved his operation east, and went all the way. Pulitzer was the first to cram
a paper with pictures and games under shrieking headlines.
He offered eight packed pages of thrilling content for only two
cents.

So Hearst was at once influenced by Pulitzer and hoping
to best him. And young Hearst was lucky enough to have a
family fortune behind his ventures. His father, George Hearst,
a rough-and-tumble Western miner and prospector, had ex-
tracted vast silver ore from the Comstock Lode in Nevada and
gold from many other mines. Young William had grown up
rich, yet he'd always been fired by ambition to achieve his own
spectacular success.

He found it in his talent for the yellow press. "HUNGRY,
FRANTIC FLAMES," read one of Hearst's most famous *Ex-
aminer* headlines, describing a fire in Monterey Bay: "Leap-
ing Higher, Higher, Higher, With Desperate Desire. Running
Madly Riotous Through Cornice, Archway and Facade." That
headline included a key reference to the newspaper itself: "The
Examiner Sends a Special Train to Monterey to Gather Full
Details of the Terrible Disaster."

That detail was telling. It was a yellow-feature trademark,
pioneered by Pulitzer and imitated by Hearst, to make the
papers themselves key parts of a story. The idea was to grow
reader loyalty for the newspaper's brand, not only by offering
astonishing content but also by making the paper seem important and powerful.

The paper must not passively report the news. The paper
must seem to make news happen.

So it was Pulitzer's pioneering *World* that young Hearst
was out to beat when he moved to New York and bought the
New York Journal in 1895. Backed by a fortune, Hearst ped-
dled the *Journal* to readers for only one cent. That forced Pu-

litzer to drop the price of the *World* to a penny as well. Their war raged back and forth, two papers constantly topping one another with wild headlines and stories that mingled startling fact with riveting fiction.

The battle was fought hardest over the rivals' desire to place their papers at the heart of the news—to seem to actually drive events. Hearst liked to boast that his reporters did the police detectives' work for them. He even took credit for starting the Spanish-American War. After the bombing of the *Maine* and the U.S. invasion of Cuba, a Hearst headline crowed: "How Do You Like the *Journal*'s War?"

And yet for all of the competitive sensationalism of the yellow press, and for all of its unabashed war-mongering, Hearst and Pulitzer were social reformers. Both Hearst's *Journal* and Pulitzer's *World* took up the cause of organized labor. The two papers were pro-immigrant. Both men were involved in Democratic Party politics.

When the Galveston disaster came along, Hearst smelled the opportunity immediately. This might be big. And he knew just whom to send.

Winifred Black was living in Denver in the fall of 1900, working there for Hearst's pioneering news wire syndicate. On September 9, vague news of a disastrous storm in Galveston, Texas, arrived by telegraph and hit the Denver paper.

Winifred had handled this kind of thing many times before. Go to the scene, do interviews, write down the situation—giving it, of course, her trademark emotional spin as "Annie Laurie," her nom de plume—put it on the wire, and go home.

So only an hour after the Galveston news arrived in Denver, Winifred was on the train to Houston. By now the veteran felt she'd seen it all. This Galveston assignment looked routine, even dull.

Her career to date had been anything but ordinary. She'd begun in hopes of becoming an actress. As a young woman, she'd appeared on the stage in road companies of tear-jerking, ripsnorting melodramas like *The Two Orphans* and *The Wages of Sin*. She'd lived in New York City then, having come far from her childhood origins, first in what were still the wild, big woods of Wisconsin, then on a farm in Illinois.

New York City: that's where theater boomed and the touring companies were cast. As a young hopeful, Winifred stood among the crowds on Broadway in the late afternoons, watching as the great sex symbol of the day, Lillian Russell, promenaded in a white feather boa and an ostrich-feather hat, with other celebrities and beauties, all the way from the Fifth Avenue Hotel on Twenty-Third Street up to Twenty-Ninth Street.

Winifred's acting career hadn't taken off. But she'd often drawn upon her talent for dress-up, make-believe, and dramatic gesture. Having moved to San Francisco and taken a job with Hearst's *Examiner,* she faked a faint in the streets of San Francisco. That was in order to write a blistering exposé of inefficiencies in the city's hospital system. Homeless and poor women, Winifred revealed, were being treated horribly at the hospital. Next, she filed a series of heartrending stories from a Hawaiian leper colony.

Soon Winifred was hustling and lying and sweet-talking her way into places where no women—and often no reporters at all—were welcome. She'd snuck onto a rowdy, smoke-filled train car where President Cleveland was traveling. When Winifred popped up from beneath a table, the president was charmed by her boldness. She scooped the competition with an exclusive presidential interview.

At a time when proper ladies didn't attend boxing matches, Winifred Black schmoozed with the prizefighters, interviewing gigantic men from Peter Jackson to Gentleman Jim Corbett to John L. Sullivan. She attended bullfights. She met the dapper gunfighter Bat Masterson, who had a notch on his gun for each man he'd killed.

For those and dozens of other high-drama stories, Annie Laurie became widely known. Time and again, Winifred arrived first on a scene, took exclusive interviews, put a high emotional spin on a story, and got it out fast over the wires before others even knew a good story was underway.

Hearst loved her. And she adored the "Big Chief," as she and his other reporters called him. He would whoop with delight at his desk when Winifred brought him a scoop. Her inside story on the horrors of the San Francisco city hospital didn't only sell papers. It led to wholesale firings there, along with many other changes for the better.

So when he invaded Pulitzer's territory in New York, Hearst naturally brought Winifred with him. The chief didn't even tell his handpicked San Francisco reporters that he'd bought a New York paper; he just sent them a telegram from that city, ordering them there. Winifred and three male colleagues boarded the cross-country train from San Francisco, wondering what was going on.

At one stop, one of the men checked the wire service and returned to the car with the startling news that the Big Chief had bought the *Journal*. If she'd known that, Winifred always said, she never would have set foot on the train. But really she knew that whatever the Big Chief wanted, she would do.

And it was back in New York that her career really took off. Hearst put Winifred on the campaign trail with his favored presidential candidate, the populist-influenced Democrat William Jennings Bryan. Winifred was in Bryan's orbit at the Democratic convention in Chicago when the candidate thrilled the crowd with his famous "Cross of Gold" speech. She spent time with Bryan and his wife, and although she liked Bryan, she later admitted she never would have voted for him (if women had been able to vote). At Hearst's behest, she nevertheless filed glowing story after glowing story on the perennially failed candidate and great orator, painting him as a savior of America.

Then Hearst sent Winifred to Utah. From there she posted a series of scathing reports on Mormon polygamy, a practice Hearst hated and wanted to end.

By 1900, Winifred Black had been all over the world for the Big Chief. She'd put the *Journal*'s huge stamp on current events while developing her uniquely emotional, first-person style as Annie Laurie. She helped make the *Journal* the most popular paper in New York, and then, through Hearst's wire syndicate, one of the most influential in the nation. And Hearst, in turn, made Annie Laurie a star.

Arriving in Houston to cover what looked like a routine flood story in Galveston, Winifred was surprised to find the city packed with what she thought of as sensation-seekers and sightseers. Her fellow newsmen had been arriving as well, and they were all frustrated and angry.

Galveston was off limits. Martial law had evidently been declared on the island, and Houston was cooperating with it. Traffic across the bay was tightly controlled. Rail tracks were wrecked, and no boat could cross without official permission. Police and the U.S. Army were everywhere, vigorously enforcing the rules.

Rumors flew. This was a disaster of historic proportions. But reporters couldn't report it. They couldn't make their way in.

Getting to Galveston, Winifred thought, would be like going to Mars. This was turning into just her kind of thing. Not routine at all, the Galveston story was an Annie Laurie specialty.

Right away, Winifred Black began using all the tricks of her trade. She never revealed how she did it—there must have been Hearst money involved—but soon she was fully disguised as a boy: red hair under a cap, face hidden under its visor. She wore work clothes and shouldered a pick. She joined a work gang heading for the island.

Even trickier: the marshal in charge of the work gang was a friend of hers. As Winifred went down a gangway with the others toward a boat to cross the bay, he was in on her ruse. So were the two huge men to her left and right. She'd made them her confederates. They were doing their best to keep her hidden between them.

Luckily, the other men in the work crew were yelling and cheering and pushing as they neared the boat. That helped Winifred stay hidden. She kept her head down as they approached sentries with crossed bayonets on the gangway. She hoped to slip by.

Suddenly there was harsh light: pitch-pine torches held by two men. One of the armed sentries caught Winifred's eye. She froze, sure she was sunk.

In response, her allies beside her in the crew started a fake fight, yelling and shoving to distract the sentries. Her friend the marshal shouted at the sentries to intervene. The sentries, ignoring Winifred, tried to restore order among the men.

So it was that, with Winifred among them, the gang of workmen boarded the flimsy boat. When it cast off into Galveston Bay, all the other reporters were back in Houston, fuming and complaining. Once again Annie Laurie would get her scoop.

What Winifred Black saw in only twenty-four hours in Galveston, and what Annie Laurie reported to the nation for William Randolph Hearst, galvanized the entire country. Without that early on-the-scene reporting, things might have gone very differently for Galveston.

Hearst's insatiable desire to boost newspaper circulation in far-off New York made news out of what had happened to people like Daisy Thorne and Chief Ketchum and Arnold Wolfram and the St. Mary's orphans and so many other Galvestonians on that awful night. And it made news out of what was happening now in the grim city of burning corpses patrolled by Major Fayling's men. The Galveston hurricane became the first national news story of the new century.

In her time on the island, Winifred never stopped observing. She smelled the stink of rotting human flesh, saw the smoke of burning corpses that hung over the island. She marveled at the sheer scale of the destruction. She reported on the "ghouls" looting the corpses. She interviewed dozens of people—ordinary Galvestonians and Police Chief Ketchum, as well as Major Fayling, whom she quickly decided to make the hero of the piece.

She collected individual storm stories, gaining a complete sense of what people had gone through on that horrific Saturday afternoon and Saturday night. She watched the corpses and body parts pulled from the pile and carted toward the pyres; she saw the grim, nauseated faces of the workers; she stared as the bodies burned.

The veteran reporter had never seen anything like it. And she was still the only reporter here.

Getting her story to the Big Chief without leaking it ahead of time, however, proved almost as tricky for Winifred Black as secreting herself onto the island. The story's value, in yellow-press terms, lay in its exclusivity. Hearst needed to place the *Journal*'s intrepid reporter right in the midst of the action. He needed to make her lonely presence in Galveston a part of the story. That kind of thing not only sold copies; it boosted the paper's brand, and it led to long-term victories in the circulation war with Pulitzer.

So after twenty-four sleepless hours of constant reporting on an emotionally devastating scene, Winifred faced her first challenge: getting off the island and back to Houston without being arrested. The veteran reporter went on wheeling and dealing. She again enlisted the marshal who had helped her sneak in; he personally took her back across the bay in a small boat. With the marshal, Winifred then traveled by pumping a handcar up the tracks above the receding water all the way to Houston.

Exhausted, deeply shaken by all she'd seen and heard, Winifred was filthy. All she wanted was to do was file her piece, take a bath, get some sleep, and get out of Texas. At 3:00 A.M., she arrived at the Houston telegraph office, carrying the only story by an outside reporter from inside Galveston.

Yet even at that hour, she found the telegraph office a mob scene. Frustrated reporters, barred access to the island, had been pulling stories out of every Galvestonian refugee they could find or invent; now they were racing to file those stories by wire. To keep the wires open until they could find a real story, some reporters had been filing whole chapters of a popular novel. Bedraggled and dirty, and trying to keep a low profile, Winifred pushed her way through the bustling crowd toward the desk of the man who received copy.

Now she had a new problem. She had to keep the very existence of her story secret until it could get safely on the wire. And she had to cut the line, getting it in ahead of the others.

Worse, if she tried to file the story directly with one of Hearst's offices, the operator and the other reporters would notice where she was filing. Her secret would be out. The other cutthroat reporters would try to slow her down while stealing her news and reporting it as their own. The story's value would disappear.

She had one hope. There was a man back in Denver, the owner of a telegraph company, who had been bugging Winifred to steer some Hearst telegraph-news business his way. In a handy coincidence, that man's last name, like Winifred's, was Black.

She quickly wrote Mr. Black the Denver telegraph operator a note. She told the man at the desk it was personal, but the real content was this: "Get my story on the wire within half an hour and you'll open a wedge on the Hearst business."

Because both names on the transmission were Black, the harried and distracted man receiving copy believed the note was indeed personal. He jumped it ahead of the line to his general manager.

The manager and the operator paid no attention to the contents. Winifred's wire went to Denver, to Mr. Black's telegraph company, not to the Hearst organization. And Mr. Black didn't tarry: within only twenty minutes, he'd sent the story from Denver to Hearst.

Shortly, to the outrage of other reporters clamoring in the Houston telegraph office, the story was in the *Journal* and on Hearst's syndication wire: an exclusive from inside Galveston.

The competition had been scooped, yet again. The "lady

journalist," as Winifred was sometimes known, had done her job.

Now for some sleep. Or so she thought.

Hearst wasn't relying only on his star reporter Annie Laurie to make the Galveston disaster, like the invasion of Cuba, another circulation booster starring his own paper. Heart-tugging reporting was important to that effort, of course—but Hearst was also prepared, as usual, to raise and spend a lot of money to place himself and the *Journal* at the center of the story.

The *Journal* had the impact and the scope to inspire donations, big and small, on a national level. Via the paper and his national wire service, Hearst immediately began raising money and organizing help for Galveston.

Hearst and his people created a relief fund. They vigorously recruited doctors, nurses, and aides, paid by the fund, to go to Houston and Galveston and save lives. They bought vast supplies of medical equipment. They identified means of getting food and water to the stranded island.

And they chartered trains to carry all that personnel and material toward the gulf. Hearst's relief trains didn't leave only from New York. They also left from San Francisco, Los Angeles, and Chicago.

Meanwhile, in New York City, Hearst hosted a glittering fundraiser, inviting the entire New York upper crust. He called the event the New York Bazaar for Galveston Orphans; it drew large sums from the wealthy New York cohort. The Big Chief himself made a personal contribution of $50,000 (almost $1.5 million in today's money).

Inspired by Hearst's big-scale relief efforts, other great industrialists and business titans began kicking in. Andrew Carnegie, for example, gave $10,000.

Ordinary people across America quickly began contributing too. Headline writers at the *Journal* naturally picked up on the more grotesque bits of Annie Laurie's report. Since the storm had opened up a graveyard, the headline read in part "Even the Graves Give Up Their Dead." That was yellow press in a nutshell.

But sensationalism went hand in glove with flat-out, first-person appeals by Annie Laurie. She directly asked her big and loyal readership, from coast to coast, to provide help for Galveston.

"But, oh, in pity's name, in America's name," Annie Laurie begged her public, "do not delay help one single instant! Send help quickly, or it will be too late!"

And in America's name, the century's first great national effort now began. Americans of every kind responded to Galveston's tragedy.

Joseph Pulitzer, Hearst's stalwart competitor in New York, was not to be outdone by his young rival in the newspaper wars. Pulitzer's *New York World* began raising contributions for Galveston too.

And Pulitzer had his own secret weapon—a very different one from Hearst's star reporter Winifred Black: Clara Barton. Founder of the American Red Cross, she commanded far greater veneration than any reporter ever could and had long been the international face of humanitarian relief. Clara Barton's arrival on the scene of any disaster had become a cause for hope.

But Miss Barton was seventy-eight now. Some within the Red Cross said she was growing too old to travel to disaster sites. Some said that like many an aging founder, she was now holding back the very organization she'd begun. Her management style was dictatorial, people claimed, unsuited to bringing the Red Cross into the twentieth century. It was time for her to retire.

Clara Barton wasn't one to give credit to others' ideas about her fitness, or her management style. Twice in her early career she'd been denied key opportunities because she was a woman. In 1851, she opened a school—the first free school in New Jersey. After growing its enrollment to over 600 students, she had found herself replaced by a man.

Then she rose up the ranks of the U.S. Patent Office in Washington, D.C. Soon she became the first-ever full-fledged female clerk—only to be busted down to copyist when some men objected to women working in government service.

It was during the Civil War that Clara Barton found what was to be her true calling, not only as a hands-on worker but also as sole, indispensable boss. Having raised money to buy and transport nursing and medical supplies for wounded Union soldiers, she finally got government and army permission to work just behind the front lines of horrifically bloody battles like Antietam and Fredericksburg.

Many found it hard to believe that this slight, unmarried woman from Massachusetts, barely five feet tall, could handle it all. Not only did she actively tend with her own hands the most grotesque injuries of that war. She also briskly managed the logistics of relief. So effective was Clara Barton in running field hospitals that she was named the official "lady in charge" of the Army of the James. Her legend began to grow.

After the war, she hit the lecture circuit, one of the biggest entertainment media of the day, competing with the theater and the opera. Barton wore black silk and discussed her relief efforts. She commanded fees as high as those paid the most famous male speakers.

She came to know other great social reformers. Among her friends and allies were the feminist Susan B. Anthony and the former slave and eloquent civil rights activist Frederick Douglass. She took up both of their causes.

Then, traveling in Europe, she helped the Swiss Red Cross set up military hospitals during the Franco-Prussian War; she was decorated by Prussia for those efforts. Learning of Europe's Geneva Convention, which protected noncombatants and set humane rules for warfare, she brought that treaty home and presented it to President Garfield. The U.S. Senate ratified the Geneva Convention in 1882, during the presidency of Chester A. Arthur.

Amid that success, Barton founded the organization first known as the U.S. Red Cross of the Geneva Convention. Clara Barton sought to establish a permanent, politically impartial, centrally organized policy of relief for wounded and devastated people, whereas prior to the Red Cross such efforts had been coordinated on the fly and then disbanded as soon as the immediate need was met.

And yet political will in the United States to support large-scale war relief had dwindled. Never again, received opinion declared, would there be a war like the Civil War. So as early as 1881, Miss Barton convinced President Arthur that hurricanes, earthquakes, floods, and fires posed threats to civil society equal to the horrors of war. The American Red Cross would make its name by serving not only in military theaters but in disaster areas as well.

By September 1900, when word of the Galveston disaster began to spread by wire, Clara Barton had ably brought massive aid to Ohio during floods, to Florida during yellow fever.

She'd traveled to Russia to feed victims of famine. She'd negotiated intensely with the Ottoman Turks to let her into the Armenian provinces after the massacres there.

In the spring of 1889, when a series of heavy rains lifted many rivers in the northeastern United States well past the flood stage, Johnstown, Pennsylvania, washed away in one of the worst floods ever to occur in the country. The Potomac too had risen, and Clara Barton left Washington with water running two feet deep on Pennsylvania Avenue to live in Johnstown for five months straight in a series of mud-filled tents. A U.S. general in charge at Johnstown called her "a poor, lone woman." Barton rejected his gallant offers of protection, and she brought about in Johnstown what was then the biggest relief effort in U.S. history.

All of that took not only unremitting compassion for victims and a highly personal touch in tending to their needs, but also a thoroughgoing operational toughness. Clara Barton left others in awe.

And yet there were the murmurings about her management style, the hints that she was too old to travel. She had to report to a board now. Money was tight, too: the American Red Cross had just completed its work in Cuba after the Spanish-American War, building hospitals and orphanages. There was competition from both the White Cross and the New York Red Cross. In 1900, Miss Barton was eager to take another important trip to shore up the reputation of the organization she'd worked tirelessly to create.

Joseph Pulitzer's *World* had offices in the building that also housed the New York offices of the Red Cross. In September 1900, via the New York Red Cross office, Pulitzer proposed a deal by telegraph to Miss Barton in Washington.

His *World* would establish a fund for Galveston donations. The paper would steer all of the money and supplies it raised exclusively to the Red Cross—if Miss Barton herself would travel to Galveston to distribute the money and supplies. And she would travel on a train handsomely outfitted, at *World* expense, escorted, of course, by a *World* correspondent. The paper would get exclusive coverage of Red Cross efforts in Galveston.

This idea suited Clara Barton. Galveston offered her an opportunity to save lives and bring comfort on a massive scale while giving the lie to her critics and competitors.

In fact, Barton was ill and feeling weak that fall. But she was used to covering up. For all of her achievements and fortitude, she'd often been depressed and sickly. Hiding her frailness, Clara Barton took a special role in the first great national event of the century. She left for Galveston on what would become her final mission for the Red Cross.

Winifred Black had succeeded in getting her exclusive story to Hearst over the wire. Now the reporter returned to her hotel room in Houston. She intended to bathe thoroughly, sleep for forty-eight hours, and catch a train back home to Denver.

But no. As she opened the door of her room, she saw piles of telegrams. They'd come from every Hearst paper and wire center throughout the country. The Big Chief had been desperately trying to reach her.

Every telegram had the same message: relief trains were heading to Houston at Hearst's behest. They were coming from New York, San Francisco, and Chicago, among other cities. They were full of doctors, nurses, and relief supplies. This influx of personnel into Houston would be huge.

Would Winifred meet all of those trains on their arrival

in Houston? In the meantime, would she get a hospital going for them, right away, in order to treat the Galveston refugees? Would she then oversee the Hearst medical effort in Houston? And still continue to file stories on it?

One foul cup of coffee later, Winifred was back at work.

Money was not a problem, at least. The Big Chief had plenty of that. Winifred quickly identified an empty Houston high school that could serve as a hospital. She hired a driver and was carried around Houston in a carriage, at high speed, from store to store. At every store she asked the price of cots. When given a price, she promised a dollar more per cot if they were delivered to the high school within one hour. Merchants hastened to comply.

By the time the first Hearst train arrived in Houston—it came from Chicago—Winifred Black had a full-scale refugee ward set up at the high school. And she had a new job: running it.

Refugees from Galveston were pouring into Houston. With boat service partially restored, people were fleeing. During her twenty-four hours inside Galveston, Winifred Black had heard and seen the worst tragedies. She'd smelled the putrefying flesh, seen the homeless children and anguished, bereaved parents. She'd interviewed the gaunt-faced workers who dug steadily through the gruesome wreckage.

But now in Houston came fallout from all of that. There were women arriving at the makeshift Hearst hospital, actively in the first stages of labor, with nowhere else to go. People showed up with awful injuries. Many were confused and bewildered; many raved in sheer madness.

There was a problem with the Hearst relief doctors, Winifred quickly discovered. They were volunteers—nobly committing their time and energy to relief. But they held widely differing, even conflicting, theories about medical treatment. There was no standard medical practice in 1900. Some of these doctors practiced homeopathy, still very popular in the late nineteenth century, based on theories about disease rejected by what the homeopaths derided as "allopaths," practitioners of mainstream techniques that homeopaths believed treated symptoms, not causes. Still other doctors on the scene were eclectic, blending allopathy and homeopathy, each in their own way. They couldn't stop bickering.

Too, the doctors hailed from different regions of the country. On that fact alone they argued tooth and nail—whose state or region was better?—when they should have been treating patients.

Morale threatened to break down still further thanks to the doctor in charge. In a pinch, Winifred had found an older Houston physician and put him in charge of running medical operations in her makeshift ward. The young, modern doctors—of every school of thought—didn't like him.

When a group of young doctors came to Winifred to complain about the plodding old-fashioned ways of the boss, she took the blame for the man's hiring. Then she read the youngsters the riot act. She reminded them of the principles of the Hippocratic oath and sent them back to work.

After that, the doctors seemed to straighten up. Soon, the thrown-together Hearst hospital in Houston for the Galveston refugees was booming. People were being treated and actually released.

And even as she ran the ward, Winifred kept filing her heartrending stories. To Annie Laurie and the Big Chief alike, no real distinction existed between reporting the many stories of human suffering and recovery, painful and uplifting alike; treating the damaged victims of the Galveston storm; and raising money for further relief. Driving this unprecedented national relief effort, yellow-press news, big money, and human compassion all played key roles.

In America's name, as Annie Laurie put it, the nation's citizens kept responding. It wasn't only the *Journal* and the *World* and magnates like Andrew Carnegie who sent money and supplies. Church after church and charity after charity, in town after town across the country, heard and read about Galveston and the relief effort and quickly raised funds and supplies. Those funds and supplies went to the Red Cross, to the Salvation Army, and to the Central Relief Committee itself in Galveston.

And at the impromptu Hearst hospital in Houston, telegrams began arriving from all over the country offering to adopt the orphaned children of Galveston. As she kept filing her moving stories as Annie Laurie, Winifred Black was overseeing an improvised national rescue effort. Its scale seemed almost to match that of the hurricane itself.

Meanwhile, in Galveston, Chief Ketchum had begun worrying about the extent of Major Lloyd Fayling's authority. The chief thought that under the necessary condition of martial law, the major was wielding too much power with too little accountability. The adjutant general of Texas, Thomas Scurry, had arrived on the island that first Tuesday with a number of militia companies. With General Scurry's administration now fully deployed, Ketchum and others on the committee determined it proper to adjust from an ad hoc emergency basis—the flamboyant civilian Lloyd Fayling, with paramilitary command of raw recruits—to a more stable procedure: a regular, experienced soldiery keeping order. That meant shifting command from Fayling to Scurry.

But Fayling had grown used to running his own organization, and running it his own way. Giving up power wasn't easy. Chief Ketchum, sensing Fayling's resistance, played it cool at first. When he met with Fayling, he began by agreeing that it was really thanks to Fayling's great efforts that the city was in such good order now. It no longer needed such a large force.

But Ketchum followed up by asserting his own authority in no uncertain terms: Bring all your men in, Ketchum told Fayling, from all parts of the city. Assemble them at the armory for inspection.

Fayling considered this a very bad idea. But he had no choice. He knew his own authority was fully subject to the committee that had created it, and to Chief Ketchum. And he remained a stickler for discipline and hierarchy. He assembled his troops.

Chief Ketchum ordered an inspection drill outside the Tremont Hotel. About 7:00 P.M., Fayling marched his troops, in military order, from the armory down to the hotel and formed them in a hollow square outside. Then followed a wait of many hours. Fayling was sure that his men, loyal to him, were growing restive, and angry with the chief. The men refrained from outright mutiny against Ketchum, according to Fayling, only because he personally reminded them of their duty to obey the orders of the civilian authority.

At last Ketchum emerged from the hotel. He ordered Fayling to return to his men and to put them through a drill, and Fayling again had no choice but to obey. Then Chief Ketchum took charge, reviewing the troops personally by giving them a series of commands; Fayling stood aside, at attention.

His men responded poorly to Ketchum's order. Fayling, steaming at attention, was forced to watch his men bumble through the drill. He was sure their errors resulted from

Ketchum's reading the orders out of an outdated Civil War drill manual.

Chief Ketchum, for his part, concluded the review simply by ordering the men to ground arms and go home. As the recruits dispersed, Ketchum told Fayling to report to General Scurry in the hotel.

Fayling could read the writing on the wall. As he entered Scurry's office, he knew he was about to be relieved of command.

So Fayling tried to turn the tables. Before Scurry could relieve him, he took the initiative. He asked the general for relief. Just for twenty-four hours, Fayling hastened to reassure the general, just in order to get some rest. This, he calculated, would make being relieved of duty his own choice and keep it temporary.

The general thanked Major Fayling for his services. He called those services "most worthy." He relieved Fayling of duty. Making no mention of any twenty-four-hour rest, he sent Fayling out of town on a mission to escort some prominent Galvestonians to Austin to take collection of $50,000 in relief funds from the governor. General Scurry took over Galveston's security.

So it was that Clara Barton's entourage, arriving in Houston after a long delay in New Orleans, was met personally by Major Fayling on behalf of the Galveston committee. Having escorted the contingent of Galvestonians to Austin, Fayling had continued to insist he was only on a brief rest but instead was told to provide Clara Barton's entourage with any assistance it might need.

Fayling took that as an instruction to create a military

honor guard, with himself in command, to welcome and assist the great humanitarian. He hustled around and "borrowed," as he put it, a Houston militia corporal's troops. He marched that guard over to the Hutchins House, the city's finest hotel. There, backed by soldiers, he persuaded the proprietor to boot out a bunch of peddlers and salesmen—"mostly Hebrew," Fayling sniffed—who had rooms on the parlor floor. The best part of the hotel was commandeered by Fayling for the use of Miss Barton and her entourage.

On September 17, when she arrived at last at the station in Houston, Clara Barton was confronted by a surprise cadre of soldiers standing at "present arms." Major Fayling greeted her with great formality and then escorted her and the Red Cross entourage, with martial ceremony, to the Hutchins House.

There seemed to be a prevailing idea that Miss Barton would inspect relief efforts on the ground in Houston, at the volunteer hospitals set up by people like Winifred Black. But no, Miss Barton said: that was not why she had come. She wanted to get into Galveston itself as quickly as possible. She wanted to relieve the many afflicted there.

The next day, however, all travel arrangements broke down. After riding the restored tracks to Texas City, the Red Cross entourage found no boat to take them across the bay. They had to spend the entire night sitting up in day-coach cars on the harbor.

Fayling consoled himself for this glitch by keeping in mind that none of it could be considered his fault. That morning, the mayor pro-tem of Houston—the "fleshy Hebrew gentleman" of Fayling's description—had relieved him of duty in arranging Miss Barton's transportation. Fayling would later recall Miss Barton's praising him for her grand reception in Houston. It was a far better reception, he reported her saying, than what she'd experienced on arriving at Johnstown in relief of that flood. She also asked him, Fayling said, to become her military aide during this mission.

Miss Barton's mind was evidently on other things. In her memoir of traveling to Galveston, she never once mentioned Lloyd Fayling. His glory days as a high official were over. Finally crossing the bay on the morning of September 15, 1900, Clara Barton and her Red Cross workers entered the disaster scene and got right to work.

CHAPTER 15

NO TONGUE CAN TELL

"Find Daisy and marry her." That was Dr. Joe Gilbert's plan when he rode toward Texas City after the hurricane. And that's just what he did.

On Thursday, September 13, Galveston saw its first wedding since the storm. They'd had a long engagement, and now Daisy Thorne and Dr. Joe, joyfully reunited, would not extend that wait by even one day. They were joined in marriage while standing on a thick slab of mud in the aisle of Grace Episcopal Church.

Daisy's wedding attire was hastily improvised: a new black skirt, a borrowed white shirtwaist, and a white hat that had been soaked in the storm; something pink had faded into it.

She didn't mind. Having experienced all of a hurricane's fury, and having survived it, she felt strongly that she'd been given a marvelous blessing.

"To have been brought so close to the infinite," is how she put it, "and to see how small finite things are."

Dr. Joe was a man of science, but he liked to say that it was Daisy's red hair—a bringer of luck—that had saved her from death, saved all the people in her room. One of those people had a contrasting theory, however. He called Daisy the bravest person he had ever known.

Arrangements for the Gilberts' first night as a married couple were as impromptu as the wedding itself. Joe and Daisy stayed with Daisy's uncle, and because that's where the entire Thorne family were staying, along with other refugees, the house was too crowded for anything like a classic wedding night.

The family offered Joe and Daisy a room of their own. The couple refused to put anyone out. That night, they slept separately. And the next morning, Daisy Gilbert left Galveston with Dr. Joe Gilbert to begin her new life in a new century.

Even more amazing than the wedding of Daisy and Joe: Galveston was beginning a new life too.

Only three weeks after that unimaginable Saturday night, help had arrived on a massive scale from every quarter. For one thing, Clara Barton had brought the American Red Cross team—with medical supplies, tubs of carbolic acid and quicklime for disinfectant, reserves of human commitment, and deep expertise.

And Miss Barton brought more than that. From a four-story warehouse borrowed from John Sealy, she began her own appeals to the nation to send clothing, food, and money. She made specific requests: now disinfectant, now light clothing, now heavier clothing, now construction material. The confidence ordinary people placed in Clara Barton and the Red Cross clarified the disaster's seriousness and inspired a public desire to give. Like Hearst's and Pulitzer's, Barton's operation became a magnet for donations.

And Miss Barton took on the Central Relief Committee that was running Galveston. That body was all male, of course, and all white, and she was used to dealing with such groups. She startled the committee by deploying local women in administrative relief positions that she intended to long outlast her own stay in Galveston. She bucked the committee's original plan to rely long-term on temporary shelters for the thousands of homeless people on the island; she demanded that the committee instead earmark funds for low-cost permanent houses quickly.

And she broke Texas codes going back to before the Civil War. She demanded that such housing be apportioned equally to people in need, without regard for race.

The committee complied with all of her demands. By the time Clara Barton left the island, on November 15, 1900, not only had the sick and injured been treated, the hungry fed, the homeless housed, the orphans adopted, and the infected cleaned, but new ideas about equality for women and African Americans had also come to Galveston. They would grow in the years to come.

Winifred Black was meanwhile running the full-fledged Galveston-refugee hospital in Houston, busy with doctors, nurses, and aides. Hearst's and Pulitzer's two big newspapers, along with American magnates and industrialists of every kind, kept raising awesome sums of money. They kept using money, power, and connections to get both cash and supplies to the island quickly. The White House, the U.S. Army, and the Texas state government were cooperating fully too. They provided both public funding and trained manpower, not only for security but also for rebuilding infrastructure, like bridges. The telephone and telegraph companies hastened to restore verbal communication; the rail companies started repairing track; electric companies worked on light and power. The U.S. Postal Service resumed mail service as early as September 12.

And thanks to both the papers and the Red Cross, ordinary people throughout the country dug deeply into their pockets. They contributed amounts that, while modest in themselves, added up to millions. Black churches in Georgia sent money, as did labor organizations in the Midwest. Galveston fundraising events were thrown from coast to coast: organ recitals, baseball games, galas. Fire companies, Sunday school classes, and every other kind of organization pitched in. People in Canada, Mexico, France, Germany, and other countries sent money too. The whole country—even much of the world—came together to bring life back to Galveston, Texas.

All of that outside help powerfully amplified the extraordinary effort that the shattered Galvestonians themselves had begun making as early as that awful first Sunday morning. Nobody looking at the island's grotesque situation in the anguished, nightmarish days following the destruction could have had any confidence in achieving even the most fundamental and urgent of tasks: removing thousands of corpses, treating the ill and injured, ensuring food and water supplies, housing the homeless. That grim prospect redefined the very concept of an overwhelming disaster.

For the scale of destruction didn't just seem unprecedented. It really was. This was truly the worst natural disaster Americans had ever seen. While death tolls would always be imperfect, it's fair to say that around 10,000 people perished in one night. And yet within about three weeks of that awful day, thanks to the ceaseless persistence of both Galveston itself and all of turn-of-the-century America, miracles of human compassion and technical prowess were being accomplished on the island.

The *Galveston Daily News* no longer amounted to a list of dead and not-dead people. The paper quickly began accepting advertising and reporting news on the city's progress toward recovery. Telegraph lines were soon humming; new telephone lines carried voices. With the repair of the central electric dynamo came trolley service. A key aspect of martial law was eased when the saloons reopened.

The port, seemingly ruined for all time only weeks before, was actually shipping cotton—slowly and occasionally at first, but then at higher and higher volumes. The U.S. Weather Bureau office in the Levy Building was up and running, with new instruments installed on the roof. The rail bridge to the mainland was rebuilt, and trains were crossing Galveston Bay.

It's true that all of this progress went on amid continued desolation. The trolley started running long before all the bodies had been removed and burned. The tracks were cleared; trolley cars brought passengers past mountains of debris still fetid with corpses. The pyres kept burning, night and day.

For temporary living quarters, the army supplied tents. They now lined the beach, thickly erected in long rows, a sprawling encampment of white canvas. Soon there were hundreds of tents housing workers helping with relief and clearing as well as many homeless Galvestonians. Visitors began calling Galveston's growing tent community "the white city on the beach."

There was a hospital tent for the beach community. There were kitchen and dining room tents. Rabbi Cohen had taken over food service for the homeless; he personally oversaw all the cooking.

It would be many months before all of those tents would be struck. Thanks to Clara Barton's insistence on quick house building, however, many people were able to leave the beach before winter.

And many people, of course, left Galveston for good.

Still, it had become clear very quickly that the city would once again defy expectations. Galveston would survive, rebuild, go on. This was an island that few, in the early days of white conquest and settlement, would have called a good spot to establish a city. And yet a city had grown here, and it had become one of the greatest in a rising nation. The spirit that built Galveston would now rebuild it.

So there was a kind of double-edged sword involved. The same rugged mood of cavalier defiance—typically Texan, quintessentially American—that had played a role in making Galveston tragically vulnerable to disaster now played a role in helping its surviving citizens come back, amazingly quickly, from the terrors they'd experienced. This was a tough, vital, and resilient people. Chastened by disaster, overcome by grief, daunted by horrifying logistics, they nevertheless began to plan ahead.

Repair of the wharves and the rebuilding of the harbor were so swift that investment money didn't instantly flee the island, as many had initially feared. Ordinary people, meanwhile—those with homes left to repair—got busy renovating them.

Ed and Annie McCullough went to work cleaning up the mud-filled home of the white man who had taken them in. Cassie Bristol, so downcast that Sunday morning by the ruined condition of her boardinghouse—her family's livelihood—did what she'd always done, ever since her husband had died at sea so long ago: she rolled up her sleeves and got to work.

"Here goes more mortgage," Cassie told her kids. Little Louise and her older sister, Lois, were wearing used clothing, donated for Galveston relief by people far away. Lois didn't like hers because it looked just like Louise's. Cassie went back into debt to rebuild her home and business. But she had regained confidence that her children wouldn't fall into poverty. She fought back.

With all that rebuilding going on—big, like the harbor infrastructure; small, like people's houses—the men of Galveston's Deep Water Committee began looking ahead. They were considering dire issues that went far beyond cleanup and rebuilding.

If Galveston was to have a realistic chance at a future once corpses were disposed of, debris removed, streets disinfected, and houses rebuilt, even more monumental tasks would have to be faced.

Denial was now at an end. Galveston lay at sea level. Disaster could strike again at any time. The city needed protection.

With that realization came a strange new period, both in the recovery effort and in the life of the island city as a whole. Galveston would again defy expectations and astonish the nation and the world.

The engineer Colonel Henry Martyn Robert, late of the Army Corps of Engineers and author of *Robert's Rules of Order,* had helped oversee the dredging of the harbor that had made Galveston one of the world's great ports. Back then, Robert had recommended building not a full-fledged seawall on land but merely a breakwater, just offshore, to break up waves as they approached the beach. Even that modest idea had been rejected in the happy days when Galveston chose to remain ignorant of danger.

Now, amid the devastation caused by the great gulf hurricane, Colonel Robert was ready to offer the commission an even bolder recommendation. He was joined in his study of the Galveston situation by two other great American engineers: Alfred Noble, who had built a breakwater on the Chicago lakefront; and H. C. Ripley, designer of Galveston's wagon bridge to the mainland. These three developed a powerful plan.

The heroic age of big engineering was just getting underway. Given the disaster that had now occurred, these men weren't about to hold back. Here was an opportunity to think bigger than almost anyone had ever thought before.

When Robert and his team came before Galveston's new government, the city's leadership had changed form. With the cleanup and disposal, the emergency committee had disbanded—but the Texas legislature had approved a revised city charter for Galveston creating a "government by commission."

Instead of twelve alderman debating every issue, a five-member board now put each member in charge of an entire area of city government. This was another Galveston "first." Other cities throughout the nation, from Houston to Boston, would soon follow suit.

The new Galveston commission set up shop not at City Hall, but in a building at Twentieth and Market. There the board heard proposals from Robert and the other two great engineers for how to protect Galveston from future rages of nature.

Robert and his colleagues laid out something truly ambitious. Galveston needed not a breakwater in the gulf but a seawall on land, they told the commission.

Made of cement, the wall must run three miles along the gulf beach, and it must rise a full seventeen feet above sea level.

That would hold back high waves, rising floods, and devastating storm surges of any future hurricanes.

A challenging proposal. Breakwaters, standing out in the water to slow and fracture waves, are one thing. A seventeen-foot wall, running along the beach itself—that's something else. There was an audacity to it, a defiance in deciding to simply shove nature back, with hard cement, at great heights, designed and crafted by modern technological prowess.

The seawall project had an aura of heady, inspiring confidence. This was a twentieth-century idea, a plan for a new kind of American future.

And yet the engineers went even farther. Recommending a long, high seawall to hold back the gulf was only the beginning. They also proposed something more astonishing and harder to achieve.

Galveston Island should be standing higher than sea level—despite the fact that a major modern city, only now beginning to recover from devastation, was already sitting on top of it.

This plan was truly audacious. And so was its budget. The engineers estimated the combined projects at $3.5 million (nearly $90 million today). Galveston, meanwhile, was broke. It was already defaulting on its bonds. No civic engineering feat exactly like this one had ever been tried before. Any plan to pick up a city and raise an island had many strikes against it.

None of that daunted Galveston's city government. While the commission leaders started wrangling tax breaks out of the Texas legislature, hawking a new bond issue to investors who were initially skeptical, and making big shows of putting up their own money, J. M. O'Rourke Construction of Denver, Colorado, began working on the project, following Colonel Robert's detailed and pioneering specifications.

They couldn't just start building the seawall. Preparations were a critical part of this project—and that meant undertaking a series of big projects within the biggest project.

Just to get building materials to the seawall site at the beach, special rail tracks had to be laid through town. By October 1902, those rails were bringing materials—which had been delivered to the island also by rail, across the bay—downtown to the beach: thousands of carloads of crushed granite, sand, cement, timber pilings, granite blocks, reinforcing steel.

At the building site along the beach, four steam pile drivers, going all at once, banged the pilings deep into the ground. The pilings were topped with planks four feet thick. And this was just the wall's foundation.

Giant wooden forms were then erected in sections sixty feet long. From rails that ran along their tops, men poured these high forms full of cement. Steel rods went into the wet cement every three feet. Once the cement had set, the forms were broken away, revealing a smooth face of high hard wall, steel-reinforced.

To prevent undermining by tides, big granite blocks and boulders were set at the toe of the wall, extending nearly thirty feet into the gulf. On the landward side, granite and gravel filled in the wall's back, giving Galveston a new, high streetscape. Soon an embankment one hundred feet wide would run nearly level with the top of the wall.

The entire three-mile run of wall proposed by Robert and his engineering team was completed in just sixteen months. Galveston, no longer lying on the beach and open to the gulf, now had a cement seawall to prevent the worst effects of hurricanes.

People came to marvel at the construction. For the thing

really did soar. Robert had designed it to curve radically, tapering as it rose from the beach. The wall showed the sea a concave face capable of tossing the highest waves back on themselves.

That sweeping white-gray form reaching toward the big Gulf Coast sky combined with its stark utilitarian purpose to make the Galveston Seawall an engineering feat to rival the ancient Wonders of the World, at once a classical fortification and a monument of austere modernism. Along its broad, granite-capped top, people dwarfed by its massive proportions bicycled, strolled, and gazed out over the great gulf toward the horizon.

Meanwhile, the really ambitious part of the project had also begun. It was time to lift up the city. Colonel Robert's engineering team had identified a 500-block area to be elevated.

Lifting a city wasn't completely unheard of. Parts of Chicago had been elevated in the middle of the nineteenth century to improve drainage and prevent infectious disease.

Still, ratcheting up many hundreds of buildings, some of them mansions, churches, schools, and office buildings, required overwhelming confidence, grand vision, and a detailed knowledge of large-scale engineering technology. For Galveston needed not only raising up but also careful sloping.

The grade at the gulf beach on the south side of town was to be set very high, nearly level with the top of the seawall. That would put the streetscape there high above the gulf. Then the grade would start running downhill, northward toward the harbor. That would direct into the bay any water that made it over the seawall.

At the beach, therefore, the island needed to go up nearly eighteen feet. Toward the bay it had to rise only about eight.

The job also required immense patience on the part of the engineers and the many workmen who made it happen—and of all of the citizens of Galveston. The idea was to elevate every building within the 500-block area, from small shacks to grand mansions to high stone public buildings. Each building would have to rise painstakingly slowly, pushed up by jacks a quarter inch at a time, until it reached the desired height. Then the buildings would sit on stilts—temporarily—and wooden boardwalks would enable people to get around while literally tons of sand were filled in below the buildings, creating the new, higher grade.

The streets themselves would also have to be opened up, of course. Gas lines and pipes underneath would be lifted, the streets then repaved at the higher grade level. Trolley tracks had to be removed from the streets first to be reinstalled when the pavement went back down at the higher grade.

It would take years of almost unbelievable civic disruption to accomplish the lifting of Galveston. At the inception of the plan, it would have been hard to believe that this whole thing could ever really happen.

And yet it was accomplished. The planners thought the plan through; then they executed it. One of the most crucial early realizations was that a project this big could only be carried out in discrete pieces, one area at a time. Galveston was lifted in rectangular sections.

Wooden dikes were laid out to define a big rectangle of the city. All the buildings in that section were slowly jacked up in preparation for the raising of the ground and supported by the stilts. Everything within that section was now to be shored up on new sand and settled down on its new grade. Streets in that area would also be lifted and then repaved. Then a new section would be laid out with the wooden dikes, and the same procedure would begin there.

One of the biggest issues the planners faced—along with lifting up entire buildings with jackscrews—was how to get more than 16 million cubic yards of sand under those raised-up buildings. The first question was where the sand should come from. The answer, the commission determined, was to dredge sand out of the shipping channel in the harbor. It always needed dredging anyway; this plan killed two birds with one stone.

So three dredging ships, built in Germany, were brought into the bay. But bringing sand into town meant disrupting the city even further. The dredgers would have to travel down a wide canal, cut right through the middle of town, running from the harbor on the north side almost all the way south to the beach.

So Galveston went ahead and dug that canal. Houses that stood in the canal's path were removed; the plan was to return the houses once the job was over and the canal refilled. When it was completed, the waterway was 200 feet wide, 20 feet deep, and more than 3 miles long. It also had two wide turning basins—cutting even farther into the streetscape—to allow the dredgers to circle and head back to the bay for more sand.

Putting a canal through a city requires further building, for traffic still has to move across town. If a canal was going to temporarily divide Galveston, a new drawbridge had to span it.

So that bridge too was quickly built. The city that had become unrecognizable on September 9, 1900, was cleaned up now. Yet as it began a process of being lifted, section by section, Galveston would remain hard to recognize for years to come.

Getting all the dredged sand under the lifted buildings represented another massive challenge. With the canal finally cut and operating and a traffic drawbridge spanning it, the three big German dredgers, loaded with tons of sand from the bay channel, started making daily trips out of the bay, down the canal, and into the city.

Once on the canal, a dredger would approach the big marked-out section of town contained within wooden dikes. All of the buildings in that section would have been jacked up, ready. People were still living in them: hence the improvised wooden boardwalks, already in place.

The dredger would anchor at a discharge station. There, a big pipe, forty-six inches in diameter, lay on a wooden gangplank that crossed the canal at the dredger's deck level.

Men on the dredger began pumping the ship's load—a slurry blend of sand and water—into the pipe. The mixture traveled quickly through the pipe, shot with force into a series of further pipes that branched out well below the jacked-up buildings on stilts. The pipes spread throughout the entire designated section of town.

As people watched from their windows and from the boardwalks, slurry pumped from the dredgers in the canal came blasting out of those pipes. It flowed into the empty spaces under the buildings, filling the space and leveling itself.

The sand from the slurry then lay there in a deep, flat pile, settling as it dried. The water from the slurry leached back into the canal and flowed back toward the bay. It might take load after load of slurry, pumped from the anchored dredges and roaring out of the pipes, to get an entire section of city filled in, eighteen feet high in some places.

But once that sand had risen high enough, there was a new grade. The nature of the island itself—essentially always a sandbar—had been altered. Section by section, the island had become higher.

The houses on it, having been raised, were now level with

the new ground. The boardwalks could be removed, as could the dikes containing the section of town.

The dredgers then started bringing sand to another section. They began pumping the slurry in there.

Not surprisingly, this process didn't always go perfectly. Kids liked to run past the spewing pipes and get covered with muck. When drying, the slurry stank; it drew flies. People slogged across the mud on wood planks and cut through other people's houses to get where they were going. Horse-drawn road traffic backed up at the drawbridge—and the bridge closed at 9:00 P.M. Trees and plants were buried deep in the salty, sterile sand; they died by the thousands.

St. Patrick's Church weighed six million pounds. It had to go up five feet. That took 700 jackscrews; the Moody mansion on Broadway took 300.

The canal that the city had dug for the dredgers kept refilling with silt. It too had to be dredged.

And paying for all of this immense effort and bizarre inconvenience fell unequally on the citizens. The city covered moving and restoring houses that had been sitting in the path of the canal. Other homeowners, however, had to pay out of their own pockets to raise their houses. If they didn't, the city could condemn the property and then either lift up the building at its own expense or simply demolish it.

And yet there were few complaints, no defaults, virtually no condemnations of property. As amazing as the engineering effort itself, the cooperation of the citizens of Galveston in a long, tedious, messy, and gargantuan effort to protect their city for the future gave testament to the horrific memory of what they'd gone through in September 1900.

That intense civic commitment reflected a new realism. People on the narrow little island were finally facing up to something they'd long denied: vulnerability to disaster. In doing so, they showed continued faith in the ability of human-kind to devise solutions to terrifying problems, to make sac-rifices in the service of implementing ambitious plans, and to manage an unpredictable future.

And that's how Galveston, Texas, was lifted up.

There had been talk at first of a mass exodus from the island. Yet many survivors of the night of horrors, stunned and exhausted on that first Sunday morning, seemingly beyond repair, would live in the rebuilt and revived city for many years to come.

Some, like Daisy Thorne, had of course been planning to leave anyway. Daisy would live a full life with Dr. Joe Gilbert, who would enjoy a long career at the University of Texas Hospital in Austin.

Among those who stayed in Galveston, Arnold Wolfram became fast friends for life with the boy he'd saved from the whirlpooling storm drain, and who had then accompanied him through the treacherous floodwaters. Cassie Bristol did rebuild her family's life at the boardinghouse. Little Louise Bristol grew up there during the long uplifting of Galveston.

Louise left school after seventh grade to help her mother at the boardinghouse. Later, as a young woman, Louise worked for the Santa Fe Railroad in Galveston, married one of the city's most successful electrical contractors, and had a daughter of her own.

She traveled to all fifty states, taking an airplane for the first time in her eighties. By then, people had long been flock-ing around Louise to hear stories of the great Galveston flood;

she was a well-regarded raconteur, a living historian in the schools.

Louise lived to experience nearly all of the amazing advances of the American Century: interstate automobile highways, jet travel, television, the moonshot. Yet she never forgot her terrifying experience, as a seven-year-old, of what nature can do to all that human aspiration.

Rabbi Cohen too went on living in Galveston for many years, continuing to serve at Congregation B'nai Israel until his retirement in 1949. Cohen's focus remained on helping others. For years he offered a familiar sight around town on his bicycle as he organized clothing and food drives. He lobbied the legislature of Texas successfully on a multitude of reforms, from raising the age of female consent to eighteen (it had been ten) to improving prison conditions.

But the rabbi became best known for spearheading what came to be called the Galveston Movement. The idea was to bring European Jews into the United States via Galveston and the Gulf Coast, instead of the more familiar—and far more crowded—ports like New York and Philadelphia. The rabbi and his organization helped immigrants settle throughout Texas and the American West. This, they hoped, would ease social tensions surrounding immigration while building up strong Jewish communities in the West.

Cohen had found a home in Galveston, and he'd then helped it survive disaster. Now he was making his island city a doorway to new homes for more than 10,000 Jewish immigrants who passed through Galveston in the early part of the twentieth century.

Henry Cohen died in 1952. He would be remembered in Galveston for many generations to come.

Winifred Black, the intrepid Hearst reporter (briefly turned

Houston hospital executive), went back to San Francisco for the Big Chief in 1906 to report on the earthquake there. Later, she brought her emotional style to a fine point when covering what became known, partly thanks to Winifred herself, as "the trial of the century." The case was as lurid as can be: Harry Thaw, cocaine-addicted son of a Pittsburgh multimillionaire, gunned down the famous New York City architect Stanford White over White's long affair with the sex symbol Evelyn Nesbit, whom Thaw had stalked and then married.

The melodrama was tailor-made for Annie Laurie's tearjerking style. Because of that and other big, popular stories of human tragedy and venality, Winifred and other "lady reporters" soon became known in the newspaper business as "sob sisters."

But Winifred always said the rest of the sob sisters seemed more like "sap sisters" to her. Few other reporters of any kind could have gotten into Galveston when it was under martial law; probably no other reporter could have whipped up a Houston hospital for the Galveston refugees and made it work.

Annie Laurie's popularity, however, was for sob-sister stories, and thanks to those articles, in the early decades of the new century Winifred became one of America's first celebrity journalists. When she died in 1936 at the age of seventy-three, she was still publishing daily; her funeral in San Francisco brought out the mayor, other dignitaries, and thousands of adoring fans. Winifred Black pioneered a certain crusading, emotional, personal style in modern national reporting— "human interest" it was called, combined with muckraking social reform—that influenced print and broadcast journalism throughout the century to come and into the next. Her trademark relentlessness was never so important as in the relief and recovery of Galveston in 1900.

Clara Barton retired from leadership of the Red Cross in

1904. Her trip to Galveston had played a critical role not only in bringing the city back from disaster but also in improving both racial and sexual equality there. And Miss Barton had proved to all detractors that she was capable of a final trip to a disaster scene. In some ways, the Galveston hurricane, which brought a great career to a climax, outdid all the other disasters Clara Barton had ever encountered.

Boyer Gonzales, having received Nell's letter advising him to stay away, came home to Galveston immediately after the storm. And yet that frustrated painter, so long bowed down by family-business cares, lacked the determined romantic certainty of a Dr. Joe Gilbert, who had known right away what he had to do: "find Daisy and marry her." When it came to Nell, Boyer remained for a time at once a steady and a noncommittal date.

But Nell hung on. The storm had shaken her deeply. She always said there was no way to adequately describe either the storm itself or the condition of the city immediately following it. Yet she didn't leave Galveston, and somehow she remained hopeful.

For something seemed to have happened to Boyer. That October of 1900, he dedicated himself to helping his brother, Alcie, get the Gonzalez mansion repaired. Then suddenly, without any warning or much discussion, Boyer sold the family business.

He stayed on as a bookkeeper—but he was done, at last, with the management pressure his hard-driving Mexican father, Thomas, had placed on him for so long, lately from beyond the grave. Boyer Gonzales had a plan now. He wanted to become a full-time painter.

With this new and dedicated approach, Boyer began meeting with success, limited at first. Some of his marine paintings—

influenced by his friend Winslow Homer, though more and more in his own style—began selling not only in Galveston but also in Boston.

In 1904 he was invited to show his work in the Texas Pavilion at the St. Louis World's Fair. Meanwhile, Boyer and Nell carried on their long companionship. And yet the artist continued to live alone in Galveston, and he continued to struggle with depression.

Then Nell took a vacation to the Catskill and Adirondack Mountains in New York State. Left abruptly alone in Galveston, Boyer came to a sudden realization.

He followed Nell northward. He met her and her sister in New York City. Out of the blue, after all those years, he asked Nell to marry him.

She quickly accepted, and the two were married in New York at the famous Little Church around the Corner at Twenty-Ninth Street and Fifth Avenue on September 21, 1907. And that's when things took off. The very day he married Nell, Boyer was accepted by the prestigious Art Students League of New York in Woodstock. Soon he and Nell were traveling in Europe, even living for a time in Florence while Boyer studied with great painters.

Returning to Galveston, the couple began spending winters in Texas and summers in Woodstock. In 1909, he and Nell had a son. Soon Boyer Gonzales became known as one of America's great painters of seascapes and marine light.

Isaac Cline spent the first weeks after the storm coming to terms with the fact that Cora was lost to him and the children. As the dead gangs carted the corpses to the pyres and the bodies slowly burned, the Cline girls still could not believe their mother was dead. They kept waiting for her to come home.

Identifying loved ones' remains, however, had quickly become a luxury none could afford. For the Galvestonian survivors, including Isaac, there was immense pain but little closure.

Yet amazingly enough, on September 30, Cora's body was found. As a gang unearthed a trove of corpses near the site of what had been Isaac Cline's house, a dead woman was found nearly intact, wearing a dress.

Her diamond engagement ring was distinctive. Cline was on the scene; he recognized and identified it.

Out of respect for the meteorologist, the workers moved the body not to a pyre but to the Lakeview Cemetery. Cora was one of the few victims of the hurricane to given burial.

Isaac Cline continued his career in the U.S. Weather Bureau. But the main weather station for the Gulf Coast was soon moved off Galveston to New Orleans, and Cline went with it. There he headed up a huge district that included the entire Gulf Coast and much of the Southwest.

In New Orleans, he predicted with his usual accuracy the Gulf Coast flooding of 1912 and 1915. He also predicted the great Mississippi flood of 1927. He wrote further articles: in some of them, he updated his understanding of hurricane behavior, based on the events that had caught him unawares in Galveston in 1900.

Joseph Cline, for his part, was transferred to Puerto Rico soon after the Galveston storm. He set up meteorological stations in remote mountains there. Later he served the bureau in the Midwest and then, moving back to Texas, in Dallas. Both Cline brothers remain well-regarded figures in the history of American weather science.

And Willis Moore, the bureau director, who did so much in 1900 to deny the reality—first the existence, and then the path—of the most destructive hurricane ever to arrive in the United States, and did so much to prevent Galvestonians from learning about it in advance, continued his career too. As director, Moore oversaw such changes in weather technology as the use of airplanes for upper atmospheric research, wireless telegraphy of weather observations to and from ships at sea, and free-rising balloon observation.

In 1913, however, Willis Moore's vaunting ambition, always powerful, got the better of him. He began waging a secret campaign to persuade President Woodrow Wilson to appoint him secretary of agriculture.

In that process, Moore used his office to browbeat subordinates into supporting his efforts. Far from subtle, the campaign only irritated and offended Wilson. Instead of being promoted into the cabinet, Moore got fired.

Only days after the hurricane hit Galveston, the U.S. War Department in Cuba responded to the terrible reality of destruction there. The department rescinded the cable ban, which Director Moore and his man in Havana, Colonel Dunwoody, had placed on weather reporting from Cuba to the United States.

Had Father Gangoite's and Julio Jover's hurricane forecasts been allowed to leave Cuba for the Gulf Coast, there's no telling how many lives could have been saved. So the War Department acted too late. But it acted.

Willis Moore, far from rethinking the virtues of his Cuban cable ban, far from acknowledging that the ban had contributed to a disastrous experience for the people of Galveston and the whole United States, responded with wounded outrage to the department's decision to lift his ban. He asked the secretary of agriculture for permission to fight back. He wanted to punish Cuba by withholding any future U.S. Weather Bureau hurricane warnings from the people of that island.

That response to the hurricane stands in stark contrast to the response of the people of Galveston. They were shocked in 1900 out of a long fog of denial and bravado. Their awful experience of loss devastated and saddened them; it might have seemed at first to ruin them. But the hurricane also toughened them.

It took a horrible tragedy—one enhanced by the arrogance of the Weather Bureau—to make Galvestonians face up to the fearful reality of their vulnerability to nature. When they did grasp that reality, they took bold, powerful, optimistic, and highly demanding steps, involving real sacrifice and inconvenience, to meet the challenges they'd finally been forced to acknowledge. With the whole people of the United States, who came together to help, Galveston worked hard every day, in the worst imaginable conditions, to bring itself back, to survive, and finally to thrive. Galveston lifted itself up both literally and figuratively. It faced head-on both the Gulf of Mexico and the new world portended by the dawn of the century.

So denial can be overcome. Galveston proves that.

But denial can also run chillingly deep.

In Havana, only one week after the hurricane, the Cuban weatherman Julio Jover went to see Willis Moore's man there, Colonel Dunwoody. Outraged by the deadly effects of the cable ban, Jover wanted to have it out with Dunwoody. Maybe in light of what had happened to Galveston, he could get the colonel to see reason.

For one thing, Jover told Dunwoody, it's just plain wrong

to censor citizens' communication, about weather or anything else. Dunwoody said he couldn't disagree, but he also asked, "Can't the government do what it pleases?" That's the kind of reasoning the Cuban was letting himself in for.

And when it came to forecasting hurricanes, Dunwoody defended his long-standing belief that, despite Cuban practices, hurricane forecasts simply can't be made. His evidence was simple. "A cyclone has just occurred in Galveston," Dunwoody declared, "which no meteorologist predicted."

It was hard to know what to say to this. Both Jover himself and Father Gangoite had indeed predicted the hurricane's arrival in Galveston, and Dunwoody had kept that information from the people there.

Jover could only state the obvious.

"That cyclone is the same one which passed over Cuba," he reminded the Colonel.

But no. Dunwoody knew better. It couldn't have been the same storm.

"No cyclone," Dunwoody told Jover, "can ever move from Florida to Galveston."

Annie McCullough lived the rest of her life in Galveston. Newly wed to Ed McCullough when the storm hit in 1900, Annie went on to become the matriarch of a large family. In 1972, when she was ninety-five, some of her younger relations interviewed her on tape. They were especially eager to get Annie on the record about her experiences of the great Galveston hurricane of 1900.

By then, Annie's sight and hearing were poor, but her memories were sharp and clear. "I got good sense," she told her interviewers. "I'm telling you the truth."

Annie told them her story. She recalled getting her roses in tubs, watching the high surf, the escape on the dray, crossing a river on Broadway. She told them what happened at the school. And as she began going over each awful moment, and thinking of the many people she knew who had died that night, it all came rushing back to her. She found herself reliving the whole event.

"I know all those people!" she said. "I'd—you'd get tired of hearing me tell it."

The family reassured her. They wanted to hear it all.

As if speaking for everybody living Galveston in September of 1900, Annie McCullough said, "The Lord knows I'm telling the truth. There ain't nobody can dispute me that went through it."

And she put the Galveston hurricane of 1900 this way: "There's no tongue," Annie McCullough said. "No tongue can tell."

ACKNOWLEDGMENTS

Thanks to the Galveston and Texas History Center of the Rosenberg Library, with special thanks to Travis Bible, Special Collections Project Coordinator, for help in navigating the Edward Weems archive, and for noting the possibilities of the Boyer Gonzalez-Nell Hertford story. Thanks to Gene Morris of the Textual Reference Branch of the National Archives. And thanks to Nikki Diller, Curator, and the staff of the Galveston County Historical Museum. Closed by the violence of Hurricane Ike in September 2008, the museum has continued to make its important archive available to researchers; happily, its exhibits and artifacts will soon reopen to the public in new space in the Galveston County Court House.

I'd like to thank Bill Hogeland, whose enthusiasm is only surpassed by his attention to detail. Thanks to my agent, Mel Berger, a steady analog hand in a vibratory digital world. And to my editor, Peter Hubbard, a guy who saw the real drama in a story that needed to be told.

A NOTE ON FURTHER READING

Two earlier narrative works for general readers recount events of the 1900 hurricane: John Edward Weems's *A Weekend in September* and Erik Larson's *Isaac's Storm*. Both served as sources for this account and helped point the way to primary sources. An excellent secondary source of a more scholarly nature is *Galveston and the 1900 Storm* by Patricia Bellis Bixel and Elizabeth Hayes Turner.

Readers who wish to dig into the hurricane's primary record will be interested in the eyewitness accounts collected in *Through a Night of Horrors,* edited by Casey Edward Greene and Shelly Henley Kelly. The book presents many compelling stories and characters, some included in this book, others beyond its scope. A trove of survivor stories, collected for the National Public Radio program "Remembering the Galveston Storm of 1900," is available online at www.npr.org/programs/lnfsound/stories/000908.stories.html. Also online is the Galveston and Texas History Center's rich collection of the 1900 hurricane oral histories, photographs, manuscripts, death lists, and so on: www.gthcenter.org/exhibits/storms/1900/index.html.

Powerful accounts fill Paul Lester's book *The Great Galveston Disaster,* a fascinating grab bag of early press reports from the scene itself. And Isaac Cline's and Joseph Cline's memoirs, covering their entire careers—*Storms, Floods, and Sunshine* and *When the Heavens Frowned,* respectively—give special attention to the events of 1900 in Galveston.

The Storm of the Century is the first book on the 1900 storm to make use of the invaluable oral histories collected in *Island of Color* (2004) by Izola Collins, which preserves the history of Galveston's African American community from Juneteenth to the post-segregation era. The book's sections on the hurricane give voice to perspectives rarely acknowledged in standard treatments. For an overall history of Galveston, Gary Cartwright's *Galveston: A History of the Island* is highly readable and informative; a better-documented source is *Galveston: A History* by David G. McComb.

On weather, weather technology, and the U.S. Weather Bureau, accessible sources include John D. Cox's *Storm Watchers*; *Divine Wind* by Kerry Emanuel; and the amazingly comprehensive collection available on the website of the library of the National Oceanic and Atmospheric Administration (www.lib.noaa.gov/), which makes available a multitude of historic documents, including reports written by both U.S. and Cuban weathermen before, during, and after 1900. Both NOAA (www.nws.noaa.gov/pa/history/) and the National Weather Service itself (www.weather.gov) offer handy histories of the service.

Complete citations for all of those sources, along with others consulted for this book, appear in the bibliography.

BIBLIOGRAPHY

ARCHIVES

Galveston County Historical Museum, Galveston, Texas.
Galveston and Texas History Center, Rosenberg Library, Galveston, Texas.
National Archives, College Park, Maryland.

SELECTED PUBLISHED WORKS

Barton, Clara. *A Story of the Red Cross: Glimpses of Field Work*. New York: D. Appleton and Co., 1904.

Bixel, Patricia Bellis, and Elizabeth Hayes Turner. *Galveston and the 1900 Storm: Catastrophe and Catalyst*. Austin: University of Texas Press, 2000.

Black, Winifred. "Rambles Through My Memories." *Good Housekeeping* 102, nos. 1–5 (January–May 1936).

Cartwright, Gary. *Galveston: A History of the Island*. Fort Worth, TX: TCU Press, 1998.

Cline, Isaac Monroe. "Special Report on the Galveston Hurricane of September 8, 1900." *NOAA History: Galveston Storm of 1900*. http://www.history.noaa.gov/stories_tales/cline2.html.

_____. *Storms, Floods and Sunshine*. New Orleans: Pelican Publishing, 1945. Reprint 2000.

Cline, Joseph Leander. *When the Heavens Frowned*. Dallas: Mathis, Van Nort & Co., 1946. Reprint, New Orleans: Pelican Publishing, 2000.

Cox, John D. *Storm Watchers: The Turbulent History of Weather Prediction from Franklin's Kite to El Niño*. Hoboken, NJ: John Wiley & Sons, 2002.

Emanuel, Kerry A. *Divine Wind: The History and Science of Hurricanes*. New York: Oxford University Press, 2005.

Garriott, E. B. "West Indian Hurricane of September 1–12, 1900." *Monthly Weather Review* 28 (September 1900).

Grade Raising: Manuscripts. Edmund R. Cheesborough Papers, 1902–1958. www.gthcenter.org/exhibits/graderaising/Manuscripts/Cheesborough/index.html

Greene, Casey Edward, and Shelly Henley Kelly, eds. *Through a Night of Horrors: Voices from the 1900 Galveston Storm.* College Station: Texas A&M University Press, 2000.

History of the National Weather Service. www.nws.noaa.gov/pa/history/.

"The Hurricane in the Gulf: Graphic Pen Picture by a Lady Passenger on the Louisiana." *Times Picayune*, September 11, 1900.

Larson, Erik. *Isaac's Storm: A Man, a Time, and the Deadliest Hurricane in History.* New York: Crown Publishers, 1999.

Lester, Paul. *The Great Galveston Disaster: Containing a Full and Thrilling Account of the Most Appalling Calamity of Modern Times, Including Vivid Descriptions of the Hurricane.* Beaver Springs, PA: American Publishing Co., 1900. Reprint, New Orleans: Pelican Publishing, 2006.

"The Louisiana's Trip: She Had a Tussle with the Hurricane in the Gulf." *Times Picayune*, September 11, 1900.

McComb, David G. *Galveston: A History.* Austin: University of Texas Press, 1986.

Nasaw, David. *The Chief: The Life of William Randolph Hearst.* Boston: Mariner Books, 2001.

"1900 Storm: Oral Histories Online." www.gthcenter.org/exhibits/storms/1900/Oralhist/index.html.

"1900 Storm: Manuscripts." http://www.gthcenter.org/exhibits/storms/1900/Manuscripts/index.html.

Ousley, Clarence. *Galveston in Nineteen Hundred.* Atlanta: William C. Chase, 1900.

Pietruska, Jamie L. "US Weather Bureau Chief Willis Moore and the Reimagination of Uncertainty in Long-Range Forecasting." *Environment and History* 17 (2011).

Rappaport, Edward N., and Jose Fernandez-Partagas. "The Deadliest Atlantic Tropical Cyclones, 1492–1996." NOAA Technical Memorandum NWS NHC 47. www.nhc.noaa.gov/pastdeadly.shtml.

"Remembering the Galveston Storm of 1900." www.npr.org/programs/lnf sound/stories/000908.stories.html.

"Rides through a Hurricane: The Louisiana's Experience in the Center of the Storm." *New York Times*, September 11, 1900.

"Seawall." www.gthcenter.org/exhibits/seawall/index.html.

Simmen, Edward. *With Bold Strokes: Boyer Gonzales, 1864–1934.* College Station: Texas A&M University Press, 1997.

"They Recall Days of Texas Badmen: Henry and Arnold Wolfram Have Seen Texas Develop from Wilderness to Empire." *Galveston Daily News*, August 27, 1939. http://familytreemaker.genealogy.com/users/s/t/e/Scharl-S-Stewart/WEBSITE-0001/UHP-0121.html.

Turner, Elizabeth Hayes. "Clara Barton and the Formation of Public Policy

in Galveston, 1900." http://www.rockarch.org/publications/confer
ences/turner.pdf.

Udias, Augustin. *Searching the Heavens and the Earth: The History of Jesuit Observatories*. Dordrecht: Springer, 2003.

Von Herrmann, C. F. "A National Weather Service Publication in Support of the Celebration of American Weather Services . . . Past, Present and Future." http://www.nws.noaa.gov/pa/history/herrmann.php.

Weems, John Edward. *A Weekend in September*. New York: Holt, 1957.

Whitnah, Donald R. *A History of the United States Weather Bureau*. Urbana: University of Illinois Press, 1961.

INDEX

ABOUT THE AUTHOR

AL ROKER is known to more than thirty million TV viewers and has won thirteen Emmy Awards, ten for his work on NBC's *Today*. He also hosts *Wake Up with Al*, a weekday morning program on the Weather Channel. A *New York Times* bestselling author, Roker lives in Manhattan with his wife, ABC News and *20/20* correspondent Deborah Roberts, and has two daughters and a son.